Volatile Compounds Formation in Specialty Beverages

Food Biotechnology and Engineering
Series Editor:
Octavio Paredes-Lopez

Volatile Compounds Formation in Specialty Beverages
Edited by Felipe Richter Reis and Caroline Mongruel Eleutério dos Santos

Native Crops in Latin America: Biochemical, Processing, and Nutraceutical Aspects
Edited by Ritva Repo-Carrasco-Valencia and Mabel C. Tomás

For more information about this series, please visit:

Volatile Compounds Formation in Specialty Beverages

Edited by
Felipe Richter Reis and
Caroline Mongruel Eleutério dos Santos

CRC Press
Taylor & Francis Group
Boca Raton   London

CRC Press is an imprint of the
Taylor & Francis Group, an **informa** business

First edition published 2022
by CRC Press
6000 Broken Sound Parkway NW, Suite 300, Boca Raton, FL 33487–2742

and by CRC Press
2 Park Square, Milton Park, Abingdon, Oxon, OX14 4RN

Library of Congress Cataloging-in-Publication Data
Names: Richter Reis, Felipe, editor. | Eleutério dos Santos, Caroline Mongruel, editor.
Title: Volatile compounds formation in specialty beverages / edited by Felipe Richter Reis, Caroline Mongruel Eleutério dos Santos.
Description: First edition. | Boca Raton : CRC Press, 2022. | Series: Food biotechnology and engineering | Includes bibliographical references and index.
Identifiers: LCCN 2021032253 (print) | LCCN 2021032254 (ebook) | ISBN 9780367631901 (hbk) | ISBN 9781032161938 (pbk) | ISBN 9781003129462 (ebk)
Subjects: LCSH: Alcoholic beverages—Flavor and odor. | Flavoring essences. | Volatile organic compounds.
Classification: LCC TP511 .V65 2022 (print) | LCC TP511 (ebook) | DDC 663/.1—dc23/eng/20211109
LC record available at https://lccn.loc.gov/2021032253
LC ebook record available at https://lccn.loc.gov/2021032254

ISBN: 978-0-367-63190-1 (hbk)
ISBN: 978-1-032-16193-8 (pbk)
ISBN: 978-1-003-12946-2 (ebk)

DOI: 10.1201/9781003129462

Typeset in Kepler Std
by Apex CoVantage, LLC

*We dedicate this book to our families
and to our friends.*

Contents

SECTION I   Distilled Beverages

SECTION II   Fermented Beverages

Preface 1

BIOTECHNOLOGY—OUTSTANDING FACTS

Agriculture started about 12,000 years ago and ever since has played a key role in food production. We look to farmers to provide the food we need but, at the same time now more than ever, to farm in a manner compatible with the preservation of the essential natural resources of the earth. Additionally, besides the remarkable positive aspects that farming has had throughout history, several undesirable consequences have been generated. The diversity of plants and animal species that inhabit the earth is decreasing. Intensified crop production has had undesirable effects on the environment (i.e., chemical contamination of groundwater, soil erosion, exhaustion of water reserves). If we do not improve the efficiency of crop production in the short term, we are likely to destroy the very resource base on which this production relies. Thus the role of so-called sustainable agriculture in the developed and underdeveloped world, where farming practices are to be modified so that food production takes place in stable ecosystems, is expected to be of strategic importance in the future, but the future has already arrived.

The biotechnology of plants is a key player in these scenarios of the 21st century; nowadays especially, molecular biotechnology is receiving increasing attention because it has the tools of innovation for the agriculture, food, chemical, and pharmaceutical industries. It provides the means to translate an understanding of and ability to modify plant development and reproduction into enhanced productivity of traditional and new products. Plant products, whether seeds, fruits, or plant components and extracts, are being produced with better functional properties and longer shelf lives, and they need to be assimilated into commercial agriculture to offer new options to small, and to more than small, industries and finally to consumers. Within these strategies is the imperative to select crops with larger proportions of edible parts as well, thus generating less waste; it is also imperative to consider the selection and development of more environmentally friendly agriculture.

Research-driven innovations for products are being developed, but the constraints of relatively long times to reach the marketplace, intellectual property rights, uncertainty of the profitability of the products, consumer acceptance, and even caution and fear with which the public may view biotechnology are tempering the momentum of all but the most determined efforts. Nevertheless, it appears uncontestable that the food biotechnology of plants and microbials will emerge as a strategic component in providing the food crops and other products required for human well-being.

FOOD BIOTECHNOLOGY AND ENGINEERING SERIES/ OCTAVIO PAREDES-LOPEZ, PHD SERIES EDITOR

The Food Biotechnology and Engineering Series aims at addressing a range of topics at the leading edge of the food biotechnology and food microbial world. In the case of foods, it includes molecular biology, genetic engineering and metabolic aspects, science, chemistry, nutrition, medical foods and health ingredients, and processing and engineering with traditional- and innovative-based approaches. The environmental aspects of producing green foods cannot be ignored. At the world level, there are foods and beverages produced by different types of microbial technologies to which this series will give attention. It will also consider genetic modifications of microbial cells to produce nutraceutical ingredients and advances in investigations of the human microbiome as related to diets.

VOLATILES COMPOUNDS FORMATION IN SPECIALTY BEVERAGES EDITORS: FELIPE RICHTER REIS AND CAROLINE MONGRUEL ELEUTÉRIO DOS SANTOS, PROFESSORS AT FEDERAL INSTITUTE OF PARANÁ, BRAZIL

As the authors indicate very clearly, this multiauthor book addresses, besides volatile compounds, the metabolic pathways and production methods of specialty beverages. The advantage is that each subject is analyzed and described by one or more experts in that field. The book is divided into two sections: distilled and fermented beverages. It contains nine chapters authored by 27 scientists and specialists; one of the distinctive features of the publication is that more than half of them are highly qualified women. The authors come from many academic institutions and organizations in six different countries on three continents. This publication is of interest to technical people involved in the quality and production of different type of beverages, governmental organizations, food science and technology staff at universities, and baccalaureate and graduate students.

Thanks are due to the editors and to all the authors for their excellent contributions to this book. Acknowledgments are also due to the editorial group of CRC Press, especially to Mr. Stephen Zollo and to Ms. Laura Piedrahita.

Preface 2

Beverages are a convenient and versatile product that may serve either to fulfill consumers' needs for hydration or as a pleasant liquor. Among the sensory attributes of beverages that drive consumer acceptability is aroma, directly influenced by the quantity and type of volatile compounds contained within them. This book addresses volatile compounds formation for specialty beverages. Initially, the general aspects of fermented and distilled beverages are presented in order to introduce the processes used for obtaining the beverages that are dealt with in the book. The book is then divided into two sections. The first, "Distilled Beverages," is dedicated to distilled beverages. It contains a chapter on cachaça, the Brazilian spirit; a chapter on Tequila, the Mexican spirit; and a chapter on whisky, from the Scotch perspective. The second section, "Fermented Beverages," is dedicated to fermented beverages. It contains various types of beverages and their various production methods: sparkling wine; cider, the traditional fermented apple beverage; kefir, the low-acid, low-alcohol beverage with its distinctive flavor; kombucha, the refreshing carbonated beverage; and dark tea, the postfermented type of tea with a unique flavor. In addition to volatile compounds, the book addresses metabolic pathways and production methods. It is certainly a valuable read for those interested in beverage science and technology.

Editors

Felipe Richter Reis is Professor at the Federal Institute of Paraná, campus Colombo and is also professor at the Graduate Program in Food Engineering, Federal University of Paraná. He earned his BSc in Food Engineering from the Federal University of Santa Catarina, Brazil, his MSc in Food Technology from the Federal University of Paraná, Brazil, and his DSc in Food Engineering from the same institution. He has published three books on food science by a world-class publisher and is continuously publishing scientific material in this field.

Caroline Mongruel Eleutério dos Santos is Professor at the Federal Institute of Paraná, campus Colombo. She earned her BSc in Food Engineering and her MSc in Food Science and Technology from the State University of Ponta Grossa, Brazil, and her DSc in Food Engineering from the Federal University of Paraná, Brazil. The theme of her DSc thesis was "Influence of Amino Acids on the Formation of Volatile Compounds in Cider." Besides teaching, she advises students of the Food Technician Course in their curricular projects and conducts practical research and extension projects in the field of Food Science and Technology.

Contributors

Chapter 1
Felipe Richter Reis (lead contributor)
Federal Institute of Paraná, campus Colombo
Colombo, PR, Brazil
Caroline Mongruel Eleutério dos Santos
Federal Institute of Paraná, campus Colombo
Colombo, PR, Brazil

Chapter 2
Maria das Graças Cardoso (lead contributor)
Chemistry Department
Federal University of Lavras
Lavras, MG, Brazil
Ana Maria de Resende Machado
Centro Federal de Educação Tecnológica de Minas
Gerais, MG, Brazil
Alex Rodrigues Silva Caetano
Chemistry Department
Federal University of Lavras
Lavras, MG, Brazil
Gabriela Aguiar Campolina
Chemistry Department
Federal University of Lavras
Lavras, MG, Brazil
David Lee Nelson
Pharmacy Department
Federal University of Minas Gerais
Belo Horizonte, MG, Brazil

Chapter 3
Mirna Estarrón-Espinosa (lead contributor)
Centro de Investigación y Asistencia en Tecnología y Diseño del Estado de Jalisco
Zapopan, Jal. Mexico
Sandra Teresita Martín del Campo
School of Engineering and Sciences
Tecnologico de Monterrey
Querétaro, Qro. Mexico

Chapter 4
Barry Harrison (lead contributor)
The Scotch Whisky Association
Edinburgh, Scotland, UK

Chapter 5
Juliane Elisa Welke
Universidade Federal do Rio Grande do Sul
Porto Alegre, RS, Brazil
Bruna Dachery
Universidade Federal do Rio Grande do Sul
Porto Alegre, RS, Brazil
Lucas Dal Magro
Universidade Federal do Rio Grande do Sul
Porto Alegre, RS, Brazil
Karolina Cardoso Hernandes
Universidade Federal do Rio Grande do Sul
Porto Alegre, RS, Brazil

Cláudia Alcaraz Zini (lead contributor)
Universidade Federal do Rio Grande do Sul
Porto Alegre, RS, Brazil

Chapter 6
Aline Alberti
Department of Food Engineering
State University of Ponta Grossa
Brazil
Alessandro Nogueira (lead contributor)
Department of Food Engineering
State University of Ponta Grossa
Brazil

Chapter 7
Maria Gabriela Cruz Pedrozo Miguel
Department of Biology
Federal University of Lavras
Lavras, MG, Brazil
Angélica Cristina de Souza
Department of Biology
Federal University of Lavras
Lavras, MG, Brazil
Débora Mara de Jesus Cassimiro
Food Science Department
Federal University of Lavras
Lavras, MG, Brazil
Disney Ribeiro Dias
Food Science Department
Federal University of Lavras
Lavras, MG, Brazil
Rosane Freitas Schwan (lead contributor)
Department of Biology
Federal University of Lavras
Lavras, MG, Brazil

Chapter 8
Jasmina Vitas (lead contributor)
Faculty of Technology Novi Sad
University of Novi Sad
Novi Sad, Republic of Serbia
Radomir Malbaša
Faculty of Technology Novi Sad
University of Novi Sad
Novi Sad, Republic of Serbia
Stefan Vukmanović
Faculty of Technology Novi Sad
University of Novi Sad
Novi Sad, Republic of Serbia

Chapter 9
Zisheng Han
Department of Food Science
Rutgers University
New Brunswick, NJ
International Joint Laboratory on Tea Chemistry and Health Effects of Ministry of Education
Anhui Agricultural University
Hefei, China
Liang Zhang (lead contributor)
International Joint Laboratory on Tea Chemistry and Health Effects of Ministry of Education
Anhui Agricultural University, Hefei, China
State Key Laboratory of Tea Plant Biology and Utilization
Anhui Agricultural University
Hefei, China

Chapter 1

Introduction to Fermented and Distilled Beverages

Felipe Richter Reis and Caroline Mongruel Eleutério dos Santos

CONTENTS

1.1 INTRODUCTION

Beverages are a convenient way of consuming nutrients, and they can be prepared by processes that either involve microorganisms or not. In the first case, the products are obtained by fermentation and the process can be complemented by a distillation step, thus yielding fermented and distilled beverages, respectively (Figure 1.1).

Fermentative processes have been used for a long time by humankind to prepare foods and beverages with pleasant sensory features, like cheese, yogurt, and vinegar. There are various types of fermentation that differ according to the microorganisms responsible for the fermentative process. In this book, fermented beverages will be dealt with from the perspective of volatile compounds formation, as during fermentation volatile compounds are generated ultimately enriching the flavor of beverages like sparkling wine, kefir, cider, kombucha, and dark tea.

DOI: 10.1201/9781003129462-1

Figure 1.1   Fermented and Distilled Beverages.

In addition, distillation, a process in which a stream rich in alcohol is selectively separated from a stream rich in water by means of the difference in their volatilities can be carried out subsequently to fermentation, in order to produce alcoholic beverages with unique sensory characteristics. Both the effect of fermentation conditions and the contribution of maturation are important to the volatile compounds profile of distilled beverages. This book will approach the volatile compounds formation in cachaça, Tequila, and whisky.

Given the lack of texts on volatile compounds formation in specialty beverages, a book dealing with such matter contributes to scientific knowledge, providing scholars, people from the industry, and enthusiasts with valuable information that can bring new insights into volatile compounds formation in fermented and distilled beverages.

1.2 REMARKABLE ASPECTS OF AND TRENDS IN THE FERMENTATION PROCESS

Fermentation refers to any process in which a product, which can be a cell or a metabolite, is obtained though the mass culture of a microorganism. The fermentation process comprises two operations: upstream and downstream. The

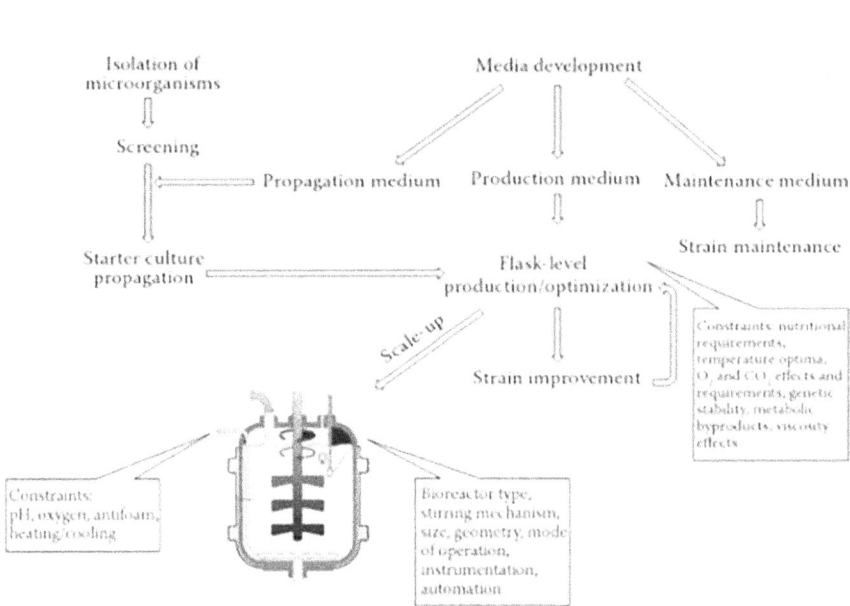

Figure 1.2   Upstream Operations in Fermentation.

Source: Reprinted with permission from Carlos Ricardo Soccol, Ashok Pandey, and Christian Larroche, *Fermentation Processes Engineering in the Food Industry* (Boca Raton: CRC Press/Taylor & Francis, 2016), 77.

operations conducted before the start of the fermenter are termed *upstream operations* (Figure 1.2) and including reactor sterilization and preparation, culture media sterilization, microbial strains, and inoculums growth and preparation. The other operations, carried out after fermentation, are called *downstream operations* (Soccol, Pandey, and Larroche 2016).

In terms of microorganisms, Alperstein *et al.* (2020) states that bioprospecting and synthetic modification of yeasts are under use for improving the quality and diversity of wine and beer. According to Cordente *et al.* (2019), there is increasing interest in *Saccharomyces cerevisiae* as a producer of aromatic compounds with industrial applications. Such compounds are synthesized from a few amino acids. Additionally, Varela (2016) affirms that there is a promising future for the development of novel non-*Saccharomyces* strains that may be used to produce wine, beer, and distilled beverages. One of these microorganisms is *Schizosaccharomyces pombe*, which has been successfully used for cider and sparkling wine production (Benito 2019).

In terms of raw materials, alternative fruit and vegetables and their combinations have been used for producing both alcoholic and nonalcoholic fermented beverages. In this sense, Santos *et al.* (2021) developed and analyzed a fermented alcoholic beverage (wine) made from *Passiflora cincinnata* mast., passion fruits, finding that this raw material is viable for wine production. The obtained wine presented high sensory scores and high antioxidant activity. In another report, Nissen, Casciano, Gianotti (2021) produced combined soy and rice beverage fermented by lactobacilli and bifidobacteria and assessed their volatile profile, finding that the mixture of raw materials was effective for producing a better flavor and reducing compounds unfavorable for the beverage aroma.

1.3 REMARKABLE ASPECTS OF AND TRENDS IN THE DISTILLATION PROCESS

Distillation is the most ancient method of separation and the most extensively used unit operation on the industry level. As shown in Figure 1.3, in *distillation*, vapor is continuously removed from the still and condensed in the condenser. The vapor is richer in the more volatile component than the liquid that remains in the still (Diwekar 1995).

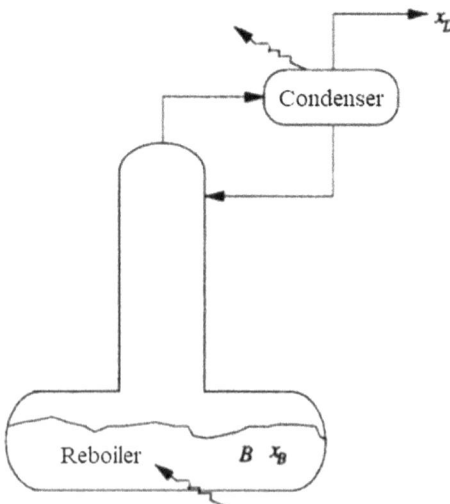

Figure 1.3   Schematic of a Simple Distillation Operation.

Source: Reprinted with permission from Umila Diwekar, *Batch Distillation: Simulation, Optimal Design and Control* (Boca Raton: CRC Press/Taylor & Francis, 1995), 2.

Regarding the volatile compounds of distilled beverages, Coldea *et al.* (2020) aged a Romanian apple brandy with different wood chips, finding that this is a suitable, less expensive alternative to aging in a wooden barrel for the formation of volatile compounds in distilled beverages. Granja-Soares *et al.* (2020) compared the effects on the odorant and sensory profile of a wine spirit of traditional aging in different types of wooden barrels to that of an alternative system composed of a stainless steel tank with dipped staves and micro-oxygenation. He concluded that the latter aging system has more of an effect on the volatiles' profile than the wood type and that the alternative system employing chestnut wood staves was superior to the traditional system in the sensory quality of the spirit.

1.4 AROMAS IN ALCOHOLIC BEVERAGES

The consumer of alcoholic beverages is highly demanding regarding product quality. An important sensory attribute analyzed at the time of consumption is aroma (Figure 1.4), which is the biggest contributor to product acceptability. Aromas are formed by a complex mixture of volatile compounds, represented by esters, aldehydes, higher alcohols, organic liquids, terpenes, and medium-length and long chain fatty acids (Januszek, Satora, and Tarko 2020; Wang *et al.* 2016).

Figure 1.4   Aroma in Alcoholic Beverages.

Some of these volatile compounds originate from the raw material, while others are the result of biochemical transformations during the fermentation process. Some factors directly influence the formation of compounds during the process, such as the yeast strain used, the temperature during fermentation, the fermentation technology, the raw material condition, the pH of the must, the fermentation time and the technology used in the maturation product (Matias-Guiu *et al.* 2020).

Esters that have a significant impact on the aroma of the final beverage are ethyl acetate, hexyl acetate (aroma of green apples), isoamyl acetate (banana flavor), ethyl lactate, ethyl butanoate, ethyl hexanoate, ethyl octanoate, ethyl decanoate, ethyl dodecanoate, ethyl hexadecanoate, ethyl-2-methylbutanoate and diethyl succinate (Buglass 2011; Fan, Xu, and Han 2011; Villière *et al.* 2012). Organic acids with the greatest aromatic impact are acetic, malic, lactic, citric, formic, fumaric, pyruvic, decanoic and octanoic acids (Zhao *et al.* 2014). Among alcohols, the most important for aromatic perception are 2-phenylethanol, phenylethyl alcohol, 1-propanol, isobutanol, 2,3-butanediol, 1-octanol, 1-decanol, 3-methyl-1-butanole 3-methylbutan-1-ol (Nicolli *et al.* 2018). Finally, other compounds like 1,1-diethoxy-ethane, 2-(4-methylcyclohex-3-en-1-yl)-propan-2-ol, furan-2-carboxaldehyde are also important.

The production routes of such compounds have been widely studied in an attempt to improve the aromatic composition of alcoholic beverages. Higher alcohols originate from corresponding amino acids via the Ehrlich metabolic pathway (Santos *et al.* 2016). Esters are derived from the raw material and obtained during the fermentation process through reactions with the alcohols present in the medium. Terpenes are formed by combining carbon chains containing five or more carbon units, leading to the formation of C10 monoterpenes, C15 sesquiterpenes, or C20 diterpenes (Januszek, Satora, and Tarko 2020).

In distilled beverages, there is also the contribution of the aging process in wooden barrels or through some other technique involving wood pieces. Spirits are aged in wooden barrels to stabilize color and improve limpidity. Additionally, there is the release of various compounds from the wood, like ellagitannins, lactones, coumarins, polysaccharides, hydrocarbons and fatty acids, terpenes, norisoprenoids, steroids, carotenoids, and furan compounds, which enrich the sensory characteristics of the product. Oak is mainly used for making the barrels, but acacia, chestnut, cherry, and mulberry are also used in some cases (De Rosso *et al.* 2009)

1.5 ROLE OF SIMPLE OR MIXED CULTURES IN THE PRODUCTION OF AROMATIC COMPOUNDS

Strains of *S. cerevisiae* are adopted for the manufacturing of alcoholic beverages because they result in less variations and greater homogeneity in processing, thus avoiding stuck fermentations and low efficiency in alcohol production

(Nogueira *et al.* 2007). However, different strains of yeast can result in desirable changes in fermented apple from a sensory viewpoint (Pietrowski *et al.* 2012).

The use of mixed yeast cultures has been identified as beneficial to the aromatic quality of fermented apple, and fermentation with a mixed inoculum leads to more effective results in the process. *Saccharomyces* (fermentative) yeasts produce higher alcohol content and increase the yield of the process. On the other hand, oxidative yeasts (or non-*Saccharomyces*) contribute to a more complex aromatic formation, resulting in fruity and floral perceptions, thus obtaining fermented apple with a more pleasant aroma (Pietrowski *et al.* 2012). However, oxidative yeasts reproduce more slowly and are more sensitive to process variations. The oxidative yeasts commonly used are *Hanseniaspora*, *Torulaspora*, and *Pichia* (Basso, Alcarde, and Portugal 2016; Hu *et al.* 2018; Wei *et al.* 2020).

1.6 ANALYTICAL METHODS FOR ASSESSING VOLATILES IN BEVERAGES

Aromas are complex compounds with different interactive properties, which may impose serious difficulties in their individual analysis. Such analysis can be performed by means of either sensory or chromatographic techniques.

A quantitative descriptive analysis is an important tool used in sensory analysis, by which it is possible to identify the aromatic perception and characteristics of the taste and appearance of a beverage. The results collected by quantitative descriptive analysis can be confirmed through analytical methods (Gürbüz, Rouseff, and Rouseff 2006; Stone *et al.* 1974).

In the past, one-dimensional gas chromatography (1D-GC) was widely used for the identification and quantification of volatiles in alcoholic beverages. However, with the advance of research in this area, it was realized that the complexity of the volatile mixture was greater than previously expected (Nicolli *et al.* 2018). Currently, the most often used analytical methods, in which it is possible to obtain results that represent the entire aromatic composition of such beverages, are gas chromatography with mass spectrometry (GC-MS), gas chromatography with olfactometry (GC-O), two-dimensional gas chromatography (mass spectrometry and the association of two-dimensional chromatography methods), and mass spectrometry with gas chromatography–olfactometry (Chen *et al.* 2018; Jeleń, Majcher, and Szwengiel 2019; Sherman *et al.* 2017). The association of methods has been increasingly used, demonstrating that possible quantification flaws, such as compound coelutions and detection after heart-cutting, are eliminated.

The steps that precede identification and quantification are important to ensure that the results represent the total volatiles in the beverage. The types of volatile extraction used most in research are solid-phase microextraction (SPME) and static headspace gas extraction (López de Lerma *et al.* 2018; Nicolli

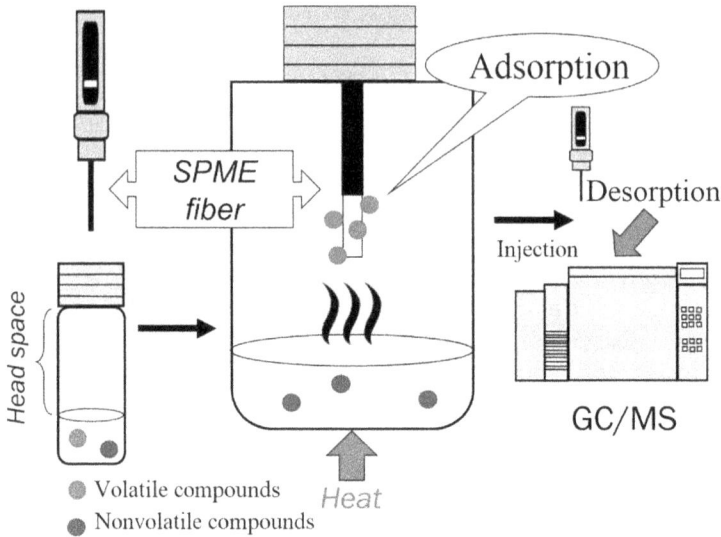

Figure 1.5   Volatile Extraction Scheme by Solid-Phase Microextraction (SPME).

et al. 2018; Wei *et al.* 2020). As shown in Figure 1.5, the fiber is exposed to the analytes in the headspace of the vial, and after the predetermined time, the analytes adsorbed on the fiber are released inside the GC column.

Comparing the methods of extraction by SPME and by static headspace, it is noticed that the use of fiber allows for a greater extraction of the volatiles present in the sample. A scheme of free volatiles (gas phase) in the headspace using the static methods is shown in Figure 1.6.

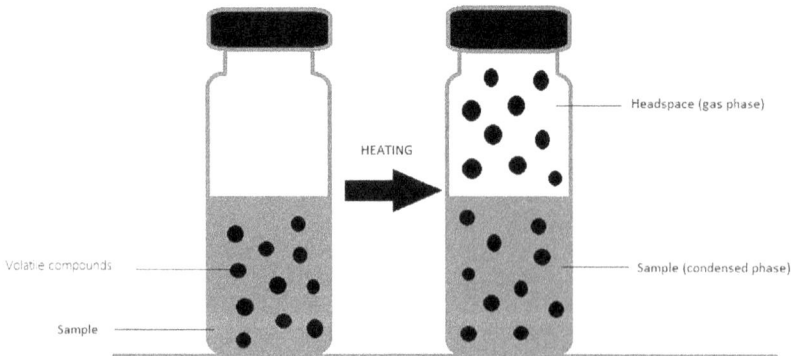

Figure 1.6   Scheme of Static Headspace Extraction.

1.7 CONCLUSIONS

Fermented and distilled beverages have been consumed for nutrition and pleas-
antness purposes for centuries. These two reasons are intrinsically connected,
as the sensory quality of foods is essential for their consumption. Volatile compounds responsible for the aroma of fermented and distilled beverages are key
components for their acceptability, to which much attention has been dedicated over the last number of years. However, this is the first attempt to gather
information about volatile compounds formation in fermented and distilled
beverages in a single document. The following chapters will provide in-depth
information with regard to such matters.

REFERENCES

Alperstein, L., J. M. Gardner, J. F. Sundstrom, J. M. Sumby, and V. Jiranek. 2020. Yeast bioprospecting versus synthetic biology—which is better for innovative beverage fermentation? *Applied Microbiology and Biotechnology* 104:1939–1953.

Basso, R. F., A. R. Alcarde, and C. B. Portugal. 2016. Could non-saccharomyces yeasts
contribute on innovative brewing fermentations? *Food Research International*
86:112–120.

Benito, S. 2019. The impacts of *Schizosaccharomyces* on winemaking. *Applied Microbiology
and Biotechnology* 103:4291–4312.

Buglass, A. J. 2011. *Handbook of Alcoholic Beverages: Technical, Analytical and Nutritional
Aspects*. Hoboken: John Wiley & Sons.

Chen, K., C. Escott, I. Loira *et al.* 2018. Use of non-saccharomyces yeasts and oenological
tannin in red winemaking: Influence on colour, aroma and sensorial properties of
young wines. *Food Microbiology* 69:51–63.

Coldea, T. E., C. Socaciu, E. Mudura *et al.* 2020. Volatile and phenolic profiles of traditional Romanian apple brandy after rapid ageing with different wood chips. *Food
Chemistry* 320:126643.

Cordente, A. G., S. Schmidt, G. Beltran, M. J. Torija, and C. D. Curtin. 2019. Harnessing
yeast metabolism of aromatic amino acids for fermented beverage bioflavouring
and bioproduction. *Applied Microbiology and Biotechnology* 103:4325–4336.

De Rosso, M., D. Cancian, A. Panighel, A. D. Vedova, and R. Flamini. 2009. Chemical compounds released from five different woods used to make barrels for aging wines
and spirits: Volatile compounds and polyphenols. *Wood Science and Technology*
43:375–385.

Diwekar, U. 1995. *Batch Distillation: Simulation, Optimal Design and Control*. Boca Raton:
CRC Press, Taylor & Francis.

Fan, W. L., Y. Xu, and Y. H. Han. 2011. Quantification of volatile compounds in Chinese
ciders by stir bar sorptive extraction (SBSE) and gas chromatography-mass spectrometry (GC-MS). *Journal of the Institute of Brewing* 117:61–66.

Granja-Soares, J., R. Roque, M. J. Cabrita *et al.* 2020. Effect of innovative technology using
staves and micro-oxygenation on the odorant and sensory profile of aged wine
spirit. *Food Chemistry* 333:127450.

Gürbüz, O., J. M. Rouseff, and R. L. Rouseff. 2006. Comparison of aroma volatiles in commercial merlot and cabernet sauvignon wines using gas chromatography– olfactometry and gas chromatography–mass spectrometry. *Journal of Agricultural and Food Chemistry* 54:3990–3996.

Hu, K., G. J. Jin, W. C. Mei, T. Li, and Y. S. Tao. 2018. Increase of medium-chain fatty acid ethyl ester content in mixed H. uvarum/S. cerevisiae fermentation leads to wine fruity aroma enhancement. *Food Chemistry* 239:495–501.

Januszek, M., P. Satora, and T. Tarko. 2020. Oenological characteristics of fermented apple musts and volatile profile of brandies obtained from different apple cultivars. *Biomolecules*. 10:853.

Jeleń, H. H., M. Majcher, and A. Szwengiel. 2019. Key odorants in peated malt whisky and its differentiation from other whisky types using profiling of flavor and volatile compounds. *LWT—Food Science and Technology* 107:56–63.

López de Lerma, N., R. A. Peinado, A. Puig-Pujol, J. C. Mauricio, J. Moreno, and T. García-Martínez. 2018. Influence of two yeast strains in free, bioimmobilized or immobilized with alginate forms on the aromatic profile of long aged sparkling wines. *Food Chemistry* 250:22–29.

Matias-Guiu, P., J. J. Rodríguez-Bencomo, J. R. Pérez-Correa, and F. López. 2020. Aroma compounds evolution in fruit spirits under different storage conditions analyzed with multiway anova and artificial neural networks. *Journal of Food Processing and Preservation* 44:14410.

Nicolli, K. P., A. C. T. Biasoto, E. A. Souza-Silva *et al.* 2018. Sensory, olfactometry and comprehensive two-dimensional gas chromatography analyses as appropriate tools to characterize the effects of vine management on wine aroma. *Food Chemistry* 243:103–117.

Nissen, L., F. Casciano, and A. Gianotti. 2021. Volatilome changes during probiotic fermentation of combined soy and rice drinks. *Food & Function* 12:3159.

Nogueira, A., C. Mongruel, D. R. S. Simões, N. Waszczynskyj, and G. Wosiacki. 2007. Effect of biomass reduction on the fermentation of cider. *Brazilian Archives of Biology and Technology*: 50.

Pietrowski, G. A. M., C. M. E. Santos, E. Sauer, G. Wosiacki, and A. Nogueira. 2012. Influence of fermentation with *Hanseniaspora* sp. yeast on the volatile profile of fermented apple. *Journal of Agricultural and Food Chemistry* 60:9815–9821.

Santos, C. M. E., A. Alberti, G. A. M. Pietrowski *et al.* 2016. Supplementation of amino acids in apple must for the standardization of volatile compounds in ciders. *Journal of the Institute of Brewing* 122:334–341.

Santos, R. T. S, A. C. T. Biasoto, A. C. P. Rybka *et al.* 2021. Physicochemical characterization, bioactive compounds, in vitro antioxidant activity, sensory profile and consumer acceptability of fermented alcoholic beverage obtained from Caatinga passion fruit (*Passiflora cincinnata* Mast.). *LWT* 148:111714.

Sherman, E., J. F. Harbertson, D. R. Greenwood, S. G. Villas-Bôas, O. Fiehn, and H. Heymann. 2017. Reference samples guide variable selection for correlation of wine sensory and volatile profiling data. *Food Chemistry* 267:344–354.

Soccol, C. R., A. Pandey, and C. Larroche. 2016. *Fermentation Processes Engineering in the Food Industry*. Boca Raton: CRC Press, Taylor & Francis.

Stone, H., J. Sidel, S. Oliver, A. Woolsey, and R. C. Singleton. 1974. *Sensory Evaluation by Quantitative Descriptive Analysis: In Descriptive Sensory Analysis in Practice*. Trumbull, CT: Food & Nutrition Press, Inc.

Varela, C. 2016. The impact of non-*Saccharomyces* yeasts in the production of alcoholic beverages. *Applied Microbiology and Biotechnology* 100:9861–9874.

Villière, A., G. Arvisenet, L. Lethuaut, C. Prost, and T. Sérot. 2012. Selection of a representative extraction method for the analysis of odourant volatile composition of French cider by GC-MS-O and GC x GC-TOF-MS. *Food Chemistry* 131:1561–1568.

Wang, J. M., D. L. Capone, K. L. Wilkinson, and D. W. Jeffery. 2016. Chemical and sensory profiles of rosé wines from Australia. *Food Chemistry* 196:682–693.

Wei, J., Y. Zhang, Y. Wang *et al.* 2020. Assessment of chemical composition and sensorial properties of ciders fermented with different non-saccharomyces yeasts in pure and mixed fermentations. *International Journal of Food Microbiology* 318:108471.

Zhao, H., F. Zhou, P. Dziugan *et al.* 2014. Development of organic acids and volatile compounds in cider during malolactic fermentation. *Czech Journal of Food Sciences* 32:69–76.

Distilled Beverages

Chapter 2

Volatile Compounds Formation in Cachaça

Maria das Graças Cardoso, Ana Maria de Resende Machado, Alex Rodrigues Silva Caetano, Gabriela Aguiar Campolina, and David Lee Nelson

CONTENTS

2.1 INTRODUCTION

Cachaça, a popular, genuinely traditional Brazilian beverage, has been conquering an increasing number of *connoisseurs* in various countries. It is the distillate most widely consumed by the population in Brazil, so the chemical composition and the volatile components must be known. These compounds act synergistically to harmonize the beverage with peculiar aromas and characteristics. Cachaça was born with the discovery of Brazil, and it has a cultural value and an economic importance that increase daily.

DOI: 10.1201/9781003129462-3

The demands of the internal and external markets cause concern with the quality of the beverage, and consequently improvements in its quality need to be implemented, not only from a commercial point of view but mainly because of the toxicological effects. A product that contains undesirable compounds can be unhealthy for the consumer.

In addition to the principal compounds, ethanol, water, and carbon dioxide, the minor compounds responsible for the bouquet of the beverage are produced in small quantities. These volatile, semivolatile or nonvolatile compounds are present in low concentrations, are formed mainly during alcoholic fermentation, and are selected by the distillation process. Such compounds belong mainly to the functional classes of acids, esters, aldehydes, and alcohols, among other derivatives of carbonyl compounds, such as acetals. Volatile compounds can be formed from derivatives of the raw material itself or through complex reactions that occur in the production process during the fermentation, distillation, storage, or aging stages. Some volatile compounds are already formed even before the fermentation process. During sugarcane grinding, volatile compounds are formed from enzymatic reactions on the polyphenols present in the sugarcane juice (Silva *et al.* 2021). During fermentation, in addition to the principal compounds, the formation of other compounds belonging to the classes of alcohols, acids, and esters occurs. A selection of volatile compounds occurs during distillation, based mainly on the boiling temperature of each compound. Other compounds can be steam distilled because of the similarity in the polarity and solubility of the compounds. In the aging stage, the contact of the beverage with the wood favors the extraction of other compounds of volatile, semivolatile, and nonvolatile natures.

The volatile constituents in cachaça are extremely important for the sensorial quality of the beverage, especially the esters and alcohols, because they are responsible for the aroma and the flavor of the beverage. In the production of a beverage, the concentrations of these compounds in the beverage should be controlled so that there is a balance in its bouquet because they can have a negative effect on the quality of the beverage and be harmful to consumer health when their concentrations are greater than the limits established by the supervisory body.

In Brazil, the supervisory body is the Ministry of Agriculture, Livestock and Supply (MAPA), which establishes acceptable limits for the presence of these secondary compounds and contaminants in sugarcane spirits and cachaça (Table 2.1) to guarantee the quality of the beverage. Normative Instruction (NI) no. 13, dated June 30, 2005 (Brazil 2005), and NI no. 28, dated August 8, 2014 (Brazil 2014), define values for the Quality and Identity Standards (QIS) of sugarcane spirits and cachaça, as described in Table 2.1. According to this Instruction, the maximum permitted concentrations of some contaminants not previously mentioned were also defined, such as ethyl carbamate (210 g L^{-1}

TABLE 2.1   STANDARDS OF IDENTITY AND QUALITY (SIQ) PRESENT IN SUGARCANE SPIRITS AND CACHAÇA.

Component	Unit	Limit	
		Minimum	Maximum
Alcohol content (spirits)	% v/v of ethyl alcohol at 20°C	38.0	54.0
Alcoholic content (cachaça)	% v/v of ethyl alcohol at 20°C	38.0	48.0
Volatile acidity in acetic acid	mg/100 mL of anhydrous alcohol	—	150.0
Esters in ethyl acetate	mg/100 mL of anhydrous alcohol	—	200.0
Aldehydes in acetic aldehyde	mg/100 mL of anhydrous alcohol	—	30.0
Furfural	mg/100 mL of anhydrous alcohol	—	5.0
Higher alcohols*	mg/100 mL of anhydrous alcohol	—	360.0
Sec-butyl alcohol (butanol-2)	mg/100 mL of anhydrous alcohol	—	10.0
Butyl alcohol (butanol-1)	mg/100 mL of anhydrous alcohol	—	3.0
Congeners**	mg/100 mL of anhydrous alcohol	200.0	650.0
Methyl alcohol	mg/100 mL of anhydrous alcohol	—	20.0
Acrolein	mg/100 mL of anhydrous alcohol	—	5.0
Ethyl carbamate	μgL^{-1}	—	210.0
Copper	mgL^{-1}	—	5.0
Arsenic	μgL^{-1}	—	100.0
Lead	μgL^{-1}	—	200.0

Source: Brasil (2005, 2014).

*Higher alcohols = (isobutyl + isoamyl + propyl).

**Congeners = (volatile acidity + esters + aldehydes + furfural + higher alcohols).

of anhydrous alcohol), acrolein (5 mg 100 mL^{-1} of anhydrous alcohol), *sec*-butyl alcohol (10 mg 100 mL^{-1} of anhydrous alcohol), butyl alcohol (3 mg 100 mL^{-1} of anhydrous alcohol), lead (200 µg L^{-1}), and arsenic (100 µg L^{-1}).

In distillation, the quality of the spirits depends fundamentally on the composition of the wine sent for distillation, the geometry of the still, or the distillation column necessary to ensure a reflux level that allows the proper separation of the secondary components, as well as the operator's ability to make the cuts at the appropriate times. In alembic distilled cachaça, depending on the degree of volatility, the distillate is divided into three fractions: head, heart, and tail. A correct separation of these fractions during the distillation contributes to the improvement of the quality of the product by minimizing the toxic metabolites (Maia and Campelo 2006). The head fraction is made up of the most volatile substances, such as methanol, acetaldehyde and ethyl acetate. It is collected in the first moments of the distillation and corresponds to 5–10% of the distillate. The head corresponds to 1% of the total wine in the still, and it has an alcohol content greater than 60°GL. The heart fraction corresponds to the real distilled spirits, representing 16% of the total volume of the still wine and 80% of the distillate. The heart fraction is more greatly appreciated, it contains less undesirable substances, and it constitutes the best portion of the distillate. The tail, the last portion of the distillate, also known as "weak water," corresponds to 3% of the total volume of the still wine (Cardoso 2020).

In column distillation, a procedure used in the production of industrial beverages, this separation of fractions does not occur, and the product is distilled in a single step. After the distillation process is finished, direct consumption is not advisable, as its quality can be improved. To add quality to the beverage, one must proceed with rest or aging. It is important that the freshly distilled beverage be stored, softened, or matured for a period of three to six months in a wooden vat, which is necessary to soften the aroma and flavor (Maia and Campelo 2006; Alcarde 2017). Aging is the last step, and it is not mandatory, but it increases the value of the beverages by improving the sensory quality. Storage is preferably accomplished in wooden barrels, where chemical reactions still take place (Maia and Campelo 2006).

There are neutral woods, such as jequitibá and peanut, that do not change the color of cachaça, and there are those that give the distillate a yellowish tone and change its aroma, such as oak, umburana, cedar, and balsam, among others. Each one lends a special touch, leaving the cachaça more or less soft, sweeter, and more fragrant, depending on the aging time (Cardoso 2020).

Alcarde *et al.* (2010) studied the aging of beverages in wooden Brazilian barrels and compared them with beverages aged in oak barrels. They found a similarity in the overall physicochemical composition of the spirits aged in the barrels of different woods. The spirits aged in oak had the greatest sensory acceptance. Among the national woods, aging in purple ipe (*Tabebuia heptaphylla*), peanut (*Pterogyne nitens*), cabreúva (*Mycrocarpus frondosus*), amburana (*Amburana cearensis*), and pear (*Platycyamus regnelli*) yielded the spirits with the best sensory qualities.

Zacaroni *et al.* (2011) determined the concentration of phenolic compounds in cachaças stored in different wooden barrels. The authors observed that the concentrations of the main extracted compounds varied according to the species of wood analyzed. The authors compared the results for beverages aged in a jatobá vat with those of beverages aged in an oak vat, and they found that the concentrations of total phenolic compounds reached 0.08 mg L^{-1} for a sample stored in a 4000–L jatobá vat over a six-month period and 40.9 mg L^{-1} for a sample aged in a 50000-L oak vat for a period of 48 months.

Santiago *et al.* (2014a) compared and quantified phenolic compounds in spirits stored in oak (*Quercus sp.*) and amburana (*Amburana cearensis*) barrels. They found that the levels of the 12 phenolic compounds analyzed increased significantly over the period of storage and that the woods furnished different principal compounds. Santiago *et al.* (2016) monitored the physicochemical quality of cachaça aged for 12 months in terms of various parameters, including the content of volatile compounds. The storage of cachaça in barrels made of oak (*Quercus sp.*), amburana (*Amburana cearensis*), jatoba (*Hymenaeae carbouril*), balm (*Myroxylon peruiferum*), and peroba (*Paratecoma peroba*) proved to be beneficial for the production of various desirable compounds. Alcohols, acids, esters and sesquiterpenes were the main groups observed.

2.2 VOLATILE CACHAÇA COMPOUNDS

2.2.1 Alcohols

Alcohols comprise the largest group of volatile substances in distilled beverages. They are formed during the oxidation process or by transformations of amino acids during the fermentation process. Ethanol is the main product of alcoholic fermentation, and, together with water, it comprises the two main components of the distillate. However, neither of these components has much influence on the flavor or bouquet of the beverage.

2.2.1.1 Methanol

Methanol, the simplest compound in the alcohol class, can also be found in the beverage. This compound is particularly undesirable, even at low concentrations. Methanol is classified by Brazilian legislation as an organic contaminant of cachaça/spirit, and its maximum limit is 20 mg 100 mL^{-1} of anhydrous alcohol (Brasil 2005).

The presence of methanol in beverages causes olfactory aggressiveness, in addition to being harmful to health. Methanol, when ingested or inhaled, undergoes an oxidation reaction in the body in the same way as ethanol. However, this reaction is slower, and it may take several days for it to be excreted. Some symptoms caused by its contamination such as headache, vertigo, vomiting, severe pain in the abdomen and blurred vision can take up to 36 hours to manifest. According to Cardoso (2020), oxidation of methanol leads to the formation of formic acid, carbon dioxide and water. Formic acid and carbon dioxide are compounds that have acidic characteristics, so they promote acidosis in the blood. Blood acidosis, when uncontrolled, can cause disturbances in the respiratory system, resulting in a coma. The most common symptom caused by methanol poisoning is visual problems; 15 mL of methanol is sufficient to cause blindness, and 10.0–100 mL can be fatal. An interesting fact is that methanol poisoning is treated with ethanol in very high doses. Ethanol competes with methanol for the enzyme alcohol dehydrogenase, thus preventing the formation of the formaldehyde and formic acid that are responsible for causing blood acidosis. Other therapeutic activities such as hemodialysis to remove formaldehyde and formic acid from the blood may be performed.

The formation of methanol occurs mostly in the fermentation stage by enzymatic or acid hydrolysis of the methoxyl group present in pectic substances. These substances are found mainly in sugarcane stalks; they are polymers of galacturonic acid with varying degrees of methoxylation. The degradation of pectic substances produces methanol and galacturonic acid, which degrades into shorter-chain sugars (Cardoso 2020).

Some measures can be taken to avoid contamination of cachaça by methanol. The separation of the head fraction must be accurate because this compound is highly volatile and most of the methanol is present in this fraction. Efficient filtration of the juice can prevent the formation of methanol during the fermentation process because methanol is formed mostly by the fermentation of ground sugarcane stalks (Evangelista 2020; Cardoso 2020).

2.2.1.2 Higher Alcohols

Higher alcohols are defined as alcohols that have more than two carbon atoms, and these alcohols are often found in distilled beverages. These compounds are known as fuel oil and generally have a floral aroma. Together with the esters, the higher alcohols are responsible for the flavor of the spirits. The principal higher alcohols found in spirits are 3-methylbutan-1-ol (isoamyl), pentan-1-ol (amyl), 2-methylpropan-1-ol (isobutyl) and propane-1-ol (propyl), in addition to significant amounts of propan-2-ol (isopropyl), butan-1-ol, (butyl), butan-2-ol (*sec*-butyl) and phenylethanol (Cardoso 2020). Higher alcohols, being volatile

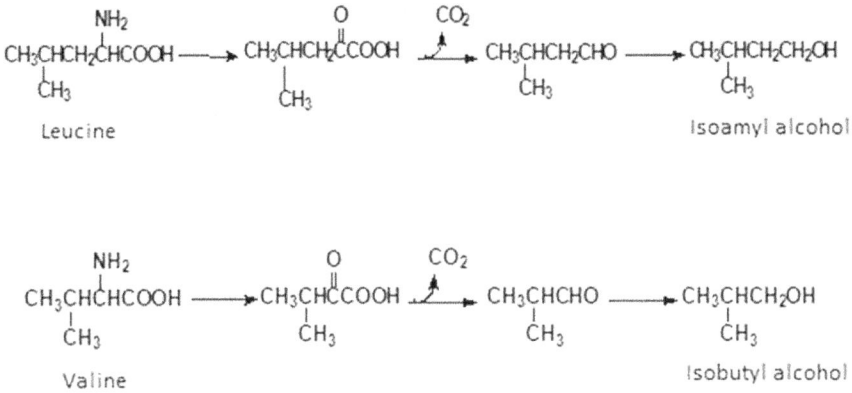

Figure 2.1   Formation of Isoamyl and Isobutyl Alcohol from the Amino Acids Leucine and Valine.

compounds, can be found mostly in the head fraction because they are carried over by hydroalcoholic vapors during distillation.

These compounds are formed in the degradation of amino acids during the fermentation process or through the degradation pathways of the sugar itself. The amino acids isoleucine, leucine, and valine initially undergo transamination catalyzed by the enzyme transaminase, forming the oxo-acid. This acid is decarboxylated with the aid of the enzyme decarboxylase to yield an aldehyde. The process can stop at this stage, or the aldehyde can be reduced to an alcohol (Figure 2.1) (Vilela *et al.* 2007; Penteado and Masini 2009).

Other factors such as species and lineage, temperature and the composition of the medium are also responsible for influencing the formation of higher alcohols by yeasts. In the fermentation stage, the amount of higher alcohols increases when the process is conducted slowly. This increase is due to the weak activity of the yeast, the high temperature and the acidic pH. The storage of sugarcane before milling can also contribute to the formation of higher alcohols (Cardoso 2020).

Soil fertilization can also be directly related to the increase in production of higher alcohols. In the work performed by Machado (2010) on the physico-chemical characterization of cachaças from sugarcane fertilized with urea, the authors found a correlation between the increase in concentrations in higher alcohols and the increase in the amount of urea in the soil. The fertilization of sugarcane with nitrogen sources can contribute to the increase in the amount of amino acids, which can result in an increase in the amount of higher alcohols in the beverage because they are formed in the fermentation stage and have their origin in amino acids.

Excess formation of higher alcohols can be avoided with some precautions during the beverage production process, such as not storing cane for a long period after cutting, thereby avoiding the degradation of amino acids and the subsequent formation of upper alcohols; not using cane that has been damage or that has been in contact with the ground; and washing the cane after cutting to prevent its contamination by bacteria that can interfere with the performance of yeasts during fermentation (Fernandes *et al.* 2007; Cardoso 2020). Other factors such as using a proprietary yeast, controlling the fermentation temperature (28–32°C), controlling the pH of the must (>4.0), avoiding excessive oxygenation in the fermentation tank and controlling the final fermentation time before distillation can also help to control the production of higher alcohols (Alcarde 2017; Bortoletto *et al.* 2018).

The production of isoamyl, amyl, isobutyl, and propyl alcohols seems to be a characteristic of yeasts in general, and the quantities produced vary with fermentation conditions, genus, species, and, probably, the strain used. These higher alcohols usually have a great influence on the flavor of alcoholic beverages. They possess characteristic aromas traditionally associated with the beverage; however, they can cause negative effects when they are present in excess, (Alcarde 2017; Cardoso 2020).

The aroma of isoamyl alcohol, for example, is characterized as "whisky" and "malt," in addition to "alcoholic," "vinyl," "banana," and "sweet." It must be remembered that these characteristics depend on the concentrations, so that very high concentrations can modify the characteristics of odors from being considered pleasant to being extremely unpleasant (Nóbrega 2003).

Santiago *et al.* (2014b) compared the physicochemical profile of cachaças aged in oak (*Quercus sp.*) and amburana (*Amburana cearensis*) barrels. The authors observed that all the samples contained desirable amounts of higher alcohols and reported that the contents differed in relation to the wood utilized.

2.2.2 Acids

Acetic acid is the main carboxylic acid formed during alcoholic fermentation. It can make up to 70% of the total acids present in yeast-distilled beverages, and it has been the main quantitative component of the acidic fraction of cachaça, expressed in volatile acidity (Cardoso 2020). In addition to this acid, lactic acid and, in smaller amounts, butyric, propionic, formic, pyruvic, citric, succinic, and oxaloacetic acids, among others, are also formed.

Acetic acid is formed through the oxidation of acetaldehyde, and its formation is stimulated by the aeration of the must and contamination by bacteria. Aeration of the must should be avoided because the increase in oxygen favors the transformation of sugar into acetic acid instead of ethanol. The contamination by acetic bacteria can occur in several stages of the production process, whether in the storage of raw material, in sugarcane juice, or even after fermentation when the distillation is not performed immediately after the fermentation process (Alcarde 2017; Cardoso 2020).

A high concentration of acids results in unpleasant sensory characteristics that directly interfere with the quality of the alcoholic beverage. In addition, they can be related to the high number of bacteria present in the fermentation of the must (Schwan and Dias 2020). However, the presence of acids in small quantities is of paramount importance because they react with alcohols to form esters, compounds that impart flavor and aroma to cachaças (Cardoso 2020).

An alternative that has been used by cachaça producers to guarantee a quality beverage for their increasingly demanding consumers has been the use of selected yeasts. These yeasts help to increase the concentration of desirable compounds during fermentation and prevent the formation of undesirable ones. Gonçalves *et al.* (2016) evaluated the chemical composition of cachaças elaborated using selected yeasts and elaborated by spontaneous fermentation. Although all the samples analyzed contained volatile compounds in accordance with the limits established by legislation, less acid was formed in the cachaças produced using selected yeasts; that is, the volatile acidity values were lower than those of cachaças produced by spontaneous fermentation.

2.2.3 Aldehydes

The aldehydes formed during alcoholic fermentation for the production of cachaça are mostly undesirable because of their toxicity, high volatility and intense odor that is detrimental to the aroma of the beverages. The ingestion of cachaças with high concentrations of these compounds can cause nasal discomfort, watery eyes, headaches, malaise, and indisposition the following day. In addition, intoxication by aldehydes can cause changes to the central nervous system; therefore, their concentrations in cachaça must be monitored (Cardoso 2020).

Acetaldehyde is a highly reactive aldehyde, it is the main aldehyde formed by the oxidation of ethanol, and it can represent up to 90% of the concentration of total aldehydes present in the beverage (Nykanen and Nykanen 1983, cited by Cardoso 2020). Its synthesis occurs mainly at the beginning of fermentation by the action of yeasts, and its production is lower under anaerobic conditions. The principal metabolic pathway involves the decarboxylation of pyruvic acid through the action of the enzyme pyruvate decarboxylase. However, acetaldehyde can also be formed by the nonenzymatic oxidation of ethanol. The concentrations of other aldehydes such as methanal, butanal, 3-methylpropanal, pentanal, hexanal, furfural, hydroxymethylfurfural, and acrolein are mainly related to the efficiency in the cuts of the head fraction during distillation because most of the aldehydes present in the wort are separated in this step.

Furfural and hydroxymethylfurfural can be formed by the decarboxylation of oxoacids or by the oxidation of alcohols, which are intermediates in the formation of these compounds (Piggott *et al.* 1989; Potter 1980, cited by Cardoso 2020). The formation of these constituents can also be linked to factors such as high temperature and low pH of the wort. These factors can cause the dehydration of sugars and the hydrolysis of the polysaccharides cellulose, hemicellulose, pectin, and others from ground sugarcane stalks.

Furfural and hydroxymethylfurfural can also be present in sugarcane juice as a result of the burning of sugarcane, which causes the contamination of cachaça and the reduction of its quality. The burning process causes dehydration and thermal degradation of sugars. Thus, furfural is formed from pentoses, and hydroxymethylfurfural is formed from hexoses. Another very important factor for controlling the formation of these aldehydes is the distillation temperature and the presence of nonvolatile compounds that can participate in the formation of furfural (Masson *et al.* 2007). Ingestion of cachaça containing concentrations of furanic aldehydes greater than the limits allowed by legislation can cause dermatitis and irritation of the mucosa and respiratory tract, in addition to problems in the central nervous system.

Rodrigues *et al.* (2020) quantified the furfural and hydroxymethylfurfural contaminants present in alembic cachaça and column cachaça in the south, midwest, and southeast regions of the states of Minas Gerais and São Paulo, Brazil, and observed that three of the 44 beverages analyzed contained concentrations of these contaminants greater than the legal limit. These three samples were obtained by column distillation. The analyses performed on beverages obtained by distillation in a copper still did not show concentrations higher than the legal limit in any sample analyzed, and the presence of furfural and hydroxymethylfurfural was not detected in any of the samples.

Propenal (acrolein) is another aldehyde that has been widely studied in recent years because of its high toxicity; it is irritating to the eyes and nose, has a spicy and bitter odor, and does not have a clear toxicity mechanism. This compound has a color that varies from colorless to yellow in its liquid form. Its formation occurs through the dehydration of glycerol during fermentation, and it can be associated with the thermoforming bacteria present in the fermentation medium (Zacaroni *et al.* 2011; Masson *et al.* 2012; Cardoso 2020).

Masson *et al.* (2012) evaluated the presence of acrolein in 71 samples of cane spirits from the northern and southern regions of the state of Minas Gerais, Brazil, and observed that the concentrations of this compound ranged from undetected to 21.97 mg 100 mL^{-1} of anhydrous alcohol; 9.85% of the samples analyzed contained concentrations greater than the maximum legal limit. In the previous year, Zacaroni *et al.* (2011) analyzed 12 samples of cachaças produced in the south of Minas Gerais, Brazil, and aged in different woods. They observed that the concentrations of acrolein ranged from undetected to 7.45 mg 100 mL^{-1} of anhydrous alcohol, and only one of the samples had a content greater than that established by law. However, the number of reports in the literature regarding the analysis of this contaminant in sugarcane spirits is still small.

2.2.4 Esters

Esters represent a class of compounds present in cachaça that, together with other secondary metabolites, are responsible for the aroma and flavor of the beverage. These components are included in one of the largest groups of compounds in the volatile fraction of alcoholic beverages. The most common esters found are ethyl acetate, ethyl formate (artificial rum flavor), isoamyl acetate (banana flavor), octyl acetate (orange), ethyl butyrate (pineapple), and pentyl butyrate (apricot) (Cardoso 2020).

The formation of aliphatic esters occurs in the secondary intracellular metabolism of yeasts during alcoholic fermentation. Another probable mechanism of ester formation involves the intermediates in the synthesis of long-chain monocarboxylic acids, where the final concentration of esters is dependent on the concentration of the alcohols and the acyl CoA produced by the yeast. Because acetyl CoA and ethanol are present in higher concentrations during fermentation, ethyl acetate is the ester present in a greater concentration than the other esters (Cardoso 2020).

In addition to the fermentative pathway, esters can be formed by esterification reactions that occur between carboxylic acids and alcohols postfermentation, during distillation, and during the aging process. However, the amount of esters formed in these stages is lower than the quantity obtained enzymatically (Santiago *et al.* 2015).

Ethyl acetate and ethyl lactate are the principal components in the parameter "total esters," the former generally being found in a greater proportion (~80%). These esters have a direct relationship with the metabolism of the yeasts themselves (*Saccharomyces cerevisiae*). In low concentrations, ethyl acetate provides a pleasant fruity aroma. However, it gives the beverage a nauseating and unwanted flavor in large concentrations. The production of ethyl lactate comes from an esterification reaction between lactic acid and ethanol. Lactic acid-producing bacteria, or lactic acid bacteria as they are also known, are present from the moment of sugarcane harvesting (Maia *et al.* 2020).

The quantification of ethyl lactate in cachaça might be an indication of peculiarities in the region of origin of the production process, but it cannot be subject to manipulation by companies. It is worth mentioning that the presence of ethyl lactate is not evidence of operational failure and that this ester, together with ethyl acetate, results exclusively from the characteristics of the broth and the fermentation and distillation stages. Unlike these esters, phenolic esters are incorporated into the beverage during storage and aging in wooden barrels (Maia *et al.* 2020).

The aroma of each ester is characteristic and depends on the alcohol portion of the molecule. The smaller the molar mass of the alcohol, the more pronounced the aroma is of the ester. Some esters such as ethyl and butyl acetate have a fruity aroma, whereas isoamyl acetate and amyl butyrate are responsible for the banana aroma. The citrus aroma is found in the acetates of alcohols with longer carbon chains. Ethyl propanoate esters, ethyl butanoate, ethyl pentanoate, ethyl heptanoate, and hexyl acetate are found in small amounts in cachaça, but they are potentially important for the aroma of the beverage because they have low odor thresholds in water (Vidal *et al.* 2013).

In addition to aliphatic esters, it is also possible to find phenolic esters of low molar mass extracted from lignin, a macromolecule in wood, and incorporated into the beverages that undergo the aging process. Compounds representing the guaiacol and seringol groups, such as ethyl syringate and ethyl vanillate, and esters extracted from wood, such as methyl homovanilate and methyl syringate, are among the main esters that have been identified in cachaça (Vichi *et al.* 2007; Castro *et al.* 2020).

Nascimento *et al.* (2008) analyzed 120 samples of cachaça and reported the generalized occurrence of ethyl lactate, pointing out its correlation with the activity of lactic bacteria present in the fermented must of the sugarcane juice. Maia *et al.* (2020) analyzed 110 brands of cachaça and found ethyl acetate and ethyl lactate in markedly varied levels. In some cases, the concentration of ethyl lactate is even greater than that of ethyl acetate.

Oliveira *et al.* (2020), when analyzing the volatile compounds of 20 cachaças produced in the state of Paraíba, Brazil, found 57 different compounds, most of which belong to the classes of alcohols and esters. Of the esters found in greatest concentrations, ethyl acetate, ethyl octanoate, ethyl decanoate, ethyl dodecanoate, ethyl tetradecanoate, and ethyl hexadecanoate stood out.

2.2.5 Ketones

Ketones are a class of organic compounds whose presence in cachaça is not limited by Brazilian legislation. The presence of ketones in cachaça is important because they contribute to the beverage's flavor. The formation of ketones occurs in secondary fermentation or by contamination during the production process. The main ketone found in alcoholic beverages is acetoin (3-hydroxybutan-2-one). This ketone is characterized by its pleasant odor, and it acts as a flavor enhancer in butter, vinegar and coffee. Acetoin is commonly found in fermentation processes promoted by yeasts and bacteria during malolactic fermentation. The acetoin route passes through diacetyl (2,3-butanedione) and

2,3-butanediol, which are important components of the flavor of various dairy products, such as cheese and butter. The reaction begins with the condensation of pyruvate with an acetaldehyde molecule, catalyzed by thiamine pyrophosphate, to form α-acetolactic acid, which subsequently undergoes oxidative decarboxylation to form acetoin (Fleet 2003).

Acetoin and 2,3-butanediol are desirable compounds in cachaça, and they are often found in wines. These two compounds do not affect the sensorial quality of alcoholic beverages at low concentration, but, in high concentrations, they can modify the bouquet and the body of the beverage because of their slightly bitter taste and viscosity.

In alcoholic beverages such as beers, wines, whisky, rum and cachaça, vicinal diketones, mainly 2,3-butanedione (diacetyl) and 2,3-pentanedione, can be found. When in excess, these compounds can cause an unpleasant taste (Rodrigues *et al.* 2020).

Several reports of cachaça have been published regarding the identification of damascenone, a ketone of the ionone group that can increase the floral-fruity aroma of the beverage, even in low concentrations. This substance can be formed after distillation through degradation of carotenoids (Prado-Jaramilo *et al.* 2015; Tábua *et al.* 2019).

2.2.6 Acetals

In rum, and possibly cachaça, most acetals are formed during distillation by the addition reaction between acetaldehyde and ethyl and pentyl alcohols to form a hemicetal, followed by condensation with another alcohol molecule (Figure 2.2). Several acetals have already been identified in cachaça (Tábua

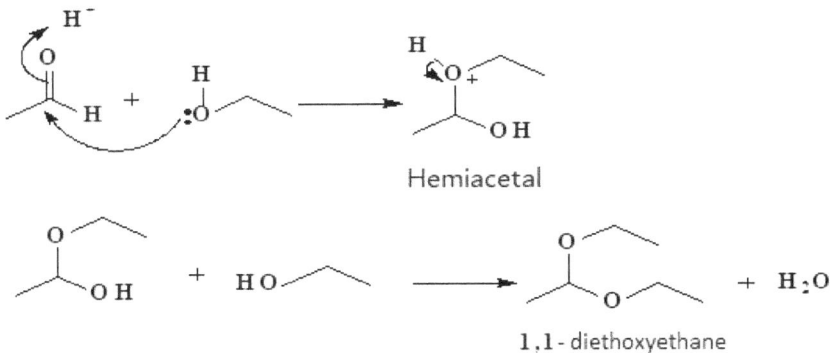

Hemiacetal

1,1- diethoxyethane

Figure 2.2   Formation of 1,1-diethoxyethane.

et al. 2019). 1,1-Diethoxyethane has a refreshing and fruity odor; therefore, it should contribute to the final aroma of cachaça, either by reducing the pungent odor of acetaldehyde, which is a major component in cachaça, or by reinforcing it with its own fruity aroma (Nóbrega 2003).

2.2.7 Ethyl Carbamate

Ethyl carbamate (EC) is considered a contaminant because of its toxicological properties, and it has been identified in several foods and beverages in which fermentation processes occur, such as wines, beers, cachaça, rum, whisky, vodka, milk, and soy derivatives, among others (Andrade-Sobrinho *et al.* 2002). In distilled fermented beverages, the formation of EC can occur before, during and after the distillation process, including during the storage of the beverage. It is identified at levels ranging from relatively low to high, relative to the MAPA value of 210 µg L^{-1} for cane spirit/cachaça (Machado *et al.* 2013; Brasil 2014; Masson *et al.* 2014).

There are currently several studies in which the differences in the EC content are correlated with the production process, mainly with the distillation and the origin of the beverage, that is, geographical differences, storage, and other such factors. The influence of the distillation and storage process on the formation of the EC in 25 samples of cane spirit from the state of Paraíba was evaluated by Nóbrega *et al.* (2009). They observed that 42% of the samples contained more than 210.0 µg L^{-1} and that the highest concentrations were in the samples distilled through columns without refrigeration. These researchers verified the existence of a close relationship between the distillation system and the formation of the EC.

Masson *et al.* (2014) detected ethyl carbamate in sugarcane spirit from the northern and southern regions of the state of Minas Gerais, Brazil. The samples were collected from small- and medium-scale alembic distillates. The concentrations determined for ethyl carbamate in the samples ranged from 23 to 980 µg L^{-1}. No correlation was observed between the levels of ethyl carbamate and the alcohol content, acidity, or copper concentrations in the samples.

Cravo *et al.* (2019) characterized the cyanogenic glycoside dhurrin in five varieties of sugarcane and determined its possible relationship with the formation of ethyl carbamate in cachaça when they investigated the genesis of ethyl carbamate in the raw material for cachaça.

Rodrigues *et al.* (2020) analyzed the contaminants in 44 samples of alembic and industrial cachaças from Minas Gerais and São Paulo and observed that two samples contained ethyl carbamate concentrations greater than the legal limits, ranging from 245.31 to 235.53 µg L^{-1}.

2.3 CONCLUSIONS

With the help of several studies that involved all the stages of the cachaça production process, a beverage that meets the sensory and physical-chemical quality parameters arrived in the consumer market. A balance between volatile, semivolatile and nonvolatile compounds in this beverage is obtained by following the recommendations of good manufacturing practices throughout the production chain.

REFERENCES

Alcarde, A. R. 2017. *Cachaça: Ciência, tecnologia e arte*. São Paulo: Ed. Blucher.

Alcarde, A. R., P. A. Souza, and A. E. S. Belluco. 2010. Aspectos da composição química e aceitação sensorial da aguardente de cana-de-açúcar envelhecida em tonéis de diferentes madeiras. *Ciência e Tecnologia de Alimentos* 30:226–232.

Andrade-Sobrinho, L. G., M. Boscolo, B. D. S. Lima-Neto, and D. W. Franco. 2002. Ethyl carbamate in alcoholic beverages (cachaça, tiquira, whisky and grape). *Química Nova* 25:1074–1077.

Bortoletto, A. M., G. C. Sivello, and A. R. Alcarde. 2018. Good manufacturing practices, hazard analysis and critical control point plan proposal for distilleries of cachaça. *Scientia Agricola* 75:432–443.

Brasil. Ministério da Agricultura, Pecuária e Abastecimento. Instrução normativa nº 13 de 29 de junho de 2005. Aprova o regulamento técnico para fixação dos padrões de identidade e qualidade para aguardente de cana e para cachaça. *Diário Oficial [da] União, Brasília*, June 30.

Brasil. Ministério da Agricultura, Pecuária e Abastecimento. Instrução normativa nº 20 de 08 de agosto de 2014. Altera o subitem 5.1.2 do Anexo da Instrução Normativa nº 13, de 29 de junho de 2005. *Diário Oficial [da] União, Brasília* Ago. 14.

Cardoso, M. G. 2020. Compostos Secundários da Cachaça. In *Produção de aguardente de cana*, edited by M. G. Cardoso, 4th ed., 445. Lavras: UFLA.

Evangelista, A. R. 2020. Aproveitamento de resíduos da fabricação de aguardente. In *Produção de aguardente de cana*, edited by M. G. Cardoso, 4th ed., 369–398. Lavras: UFLA.

Castro, M. C., A. M. Bortoletto, G. C. Silvello, and A. R. Alcarde. 2020. Lignin-derived phenolic compounds in cachaça aged in new barrels made from two oak species. *Heliyon*:e055806.

Cravo, F. D. C., W. D. Santiago, A. L. Silva *et al.* 2019. Composition of cachaças produced from five varieties of sugarcane and the correlation of the presence of dhurrin in the cane with that of ethyl carbamate in the product. *American Journal of Plant Sciences* 10:339–350.

Fernandes, W. J., M. G. Cardoso, F. J. Vilela, A. R. Morais, V. de Fátima Silva, and D. L. Nelson. 2007. Physicochemical quality of a blend of domestic cachaças from the South of Minas Gerais. *Journal of Food Composition and Analysis* 20:257–261.

Fleet, G. H. 2003. Yeast interactions and wine flavour. *International Journal of Food Microbiology* 86:11–22.

Gonçalves, R. C. F., M. M. G. Teodoro, A. M. R. Machado, F. C. O. Gomes, F. Badotti, and M. G. Cardoso. 2016. Compostos voláteis em cachaças de alambique produzidas por leveduras selecionadas e por fermentação espontânea. *Magistra* 28:285–293.

Machado, A. M. R. 2010. *Carbamato de etila, acroleína e hidrocarbonetos policíclicos aromáticos: Caracterização e quantificação em cachaças provenientes de cana-deaçúcar com adubação nitrogenada e acondicionadas em vidros e" bombonas" de Pead.* PhD Thesis. Lavras: Federal University of Lavras.

Machado, A. M. R., M. G. Cardoso, A. A. Saczk *et al.* 2013. Determination of ethyl carbamate in cachaça produced from copper stills by HPLC. *Food Chemistry* 138:1233–1238.

Maia, A. B., L. S. Marinho, D. L. Nelson. 2020. Advance in the characterization of alambic cachaça: Ethyl lactate. *Research, Society and Development* 9:e297997116.

Maia, A. B. R. A., and E. P. Campelo. 2006. *Tecnologia da Cachaça de Alambique.* Belo Horizonte, Brasil: Sebrae-SindBebidas.

Masson, J., M. G. Cardoso, F. J. Vilela, F. A. Pimentel, A. R. D. Morais, and J. P. D. Anjos. 2007. Parâmetros físico-químicos e cromatográficos em aguardentes de cana queimada e não queimada. *Ciência e Agrotecnologia* 31:1805–1810.

Masson, J., M. G. Cardoso, L. M. Zacaroni *et al.* 2012. Determination of acrolein, ethanol, volatile acidity, and copper in different samples of sugarcane spirits. *Ciência e Tecnologia de Alimentos* 32:568–572.

Masson, J., M. G. Cardoso, L. M. Zacaroni *et al.* 2014. GC-MS analysis of ethyl carbamate in distilled sugar cane spirits from the northern and southern regions of Minas Gerais. *Journal of the Institute of Brewing* 120:516–520.

Nascimento, E. S. P., D. R. Cardoso, and D. P. Franco. 2008. Quantitative ester analysis in cachaça and distilled spirits by gas chromatography-mass spectrometry (CG-MS). *Journal of Agricultural and Food Chemistry* 56:5488–5493.

Nóbrega, I. C. C. 2003. Análise dos compostos voláteis da aguardente de cana por concentração dinâmica do "headspace" e cromatografia gasosa-espectrometria de massas. *Ciência e Tecnologia de Alimentos* 23:210–216.

Nóbrega, I. C. C., J. A. Pereira, J. E. Paiva, and D. W. Lachenmeier. 2009. Ethyl carbamate in pot still cachaças (Brazilian sugar cane spirits): Influence of distillation and storage conditions. *Food Chemistry* 117:693–697.

Nykanen, L., and I. Nykanen. 1983. *Rum flavour of distilled beverages: Origin and development.* Chichester: E. Harwood.

Oliveira, R. E., M. G. Cardoso, W. D. Santiago, R. B. Barbosa, G. F. Alvarenga, and D. L. Nelson. 2020. Physicochemical parameters and volatile composition of cachaça produced in the state of Paraíba, Brasil. *Research, Society and Development* 9: e504974409.

Penteado, J. C. P., and J. C. Masini. 2009. Heterogeneidade de álcoois secundários em aguardentes brasileiras de diversas origens e processos de fabricação. *Química Nova* 32:1212–1215.

Piggott, J. K. *et al.* 1989. *The science and technology of whiskies.* New York: Longman.

Potter, N. N. 1980. *Food science.* Westport: Avi, 653.

Prado-Jaramilo, N., M. Estarrón-Espinosa, H. Escalona-Buendía, R. Cosío-Ramírez, and S. T. Martíndel-Campo. 2015. Volatile compounds generation during different stages of the Tequila production process: A preliminary study. *LWT- Food Science and Technology* 61:471–483.

Rodrigues, L. M. A., M. G. Cardoso, W. D. Santiago *et al*. 2020. Organic contaminants in distilled sugar cane spirits produced by column and copper alembic distillation. *Research, Society and Development* 9: e930974879.

Santiago, W. D., M. G. Cardoso, M. S. Gomes, L. M. A. Rodrigues, R. R. Cardoso, and R. M. Brandão. 2014a. Correlação entre extrato seco total, composição fenólica total e intensidade de cor de cachaças envelhecidas em tonéis de carvalho (*Quercus* sp.) e amburana (*Amburana cearensis*) em um período de 12 meses. *E-xacta* 7:9–15.

Santiago, W. D., M. G. Cardoso, J. A. Santiago *et al*. 2014b. Comparison and quantification of the development of phenolic compounds during the aging of cachaça in oak (*Quercus* sp.) and amburana (*Amburana cearensis*) barreis. *American Journal of Plant Sciences* 5:3140–3150.

Santiago, W. D., M. G. Cardoso, J. A. Santiago *et al*. 2016. Physicochemical profile and determination of volatile compounds in cachaça stored in new oak (*Quercus* sp.), amburana (*Amburana* cearensis), jatoba (*Hymenaeae carbouril*), balsam (*Myroxylon peruiferum*) and peroba (*Paratecoma peroba*) casks by SPME-GC-MS. *Journal of the Institute of Brewing* 122:624–634.

Santiago, W. D., M. G. Cardoso, L. M. Zacaroni *et al*. 2015. Multivariate analysis for the characterization of physico-chemical profiles of cachaça produced in copper stills over a period of six years in Minas Gerais state. *Journal of the Institute of Brewing* 121:244–250.

Schwan, R. F., and D. R. Dias. 2020. Fermentação. In *Produção de aguardente de cana*, edited by M. G. Cardoso, 4th ed., 93–117. Lavras: UFLA.

Silva, P., J. Freitas, F. M. Nunes, and J. S. Câmara. 2021. Effect of processing and storage on the volatile profile of sugarcane honey: A four-year study. *Food Chemistry* 365:130457.

Tábua, M. C. M., W. D. Santiago, M. L. Magalhães *et al*. 2019. Identification of volatile compounds, quantification of glycerol and trace elements in distilled spirits produced in Mozambique. *Journal of Food Science and Technology* 57:505–512.

Vichi, S., C. Santini, N. Natali, C. Riponi, E. Lopez-Tamames, and S. Buxaderas. 2007. Volatile and semi-volatile components of oak wood chips analysed by accelerated solvent extraction coupled to gas chromatography—mass spectrometry (GC-MS). *Food Chemistry* 102:1260–1269.

Vidal, E. E., G. M. Billerbeck, D. A. Simões, A. Schuler, J. M. François, and M. A. Morais Jr. 2013. Influence of nitrogen supply on the production of higher alcohols/esters and expression of flavour-related genes in cachaça fermentation. *Food Chemistry* 138:701–708.

Vilela, F. J. V., M. G. Cardoso, J. Masson, and J. P. Anjos. 2007. Determinação das composições físico-químicas de cachaça do Sul de Minas Gerais e de suas misturas. *Ciência e Agrotecnologia* 31:1089–1094.

Zacaroni, L. M., M. G. Cardoso, A. A. Saczk *et al*. 2011. Caracterização e quantificação de contaminantes em aguardentes de cana. *Química Nova* 34:320–324.

Chapter 3

Volatile Compounds Formation in Tequila

Mirna Estarrón-Espinosa and Sandra Teresita Martín del Campo

CONTENTS

3.1 INTRODUCTION

Tequila is a typical Mexican alcoholic beverage. It is obtained from a double distillation of fermented wort, prepared with sugars extracted from *Agave tequilana* Weber var. Azul. This beverage can be elaborated only in the Denomination of Origin Region (DOT) with *A. tequilana* cultivated into it. The DOT includes the Mexican territory delimited by the state of Jalisco, as well as some municipalities in the states of Nayarit, Michoacán, Guanajuato, and Tamaulipas, and is referred to in the Official Standard (NOM-006-SCFI-2012 2012).

There are two categories of this distilled alcoholic beverage: One is 100% Agave Tequila, made using sugars derived exclusively from the hydrolyzed juices of *A. tequilana* Weber var. Azul. The other is Tequila made from wort prepared with 51% of sugars from *A. tequilana* Weber var. Azul and a proportion no higher than 49% of other sources such as cane sugar, cane molasses, and

DOI: 10.1201/9781003129462-4

33

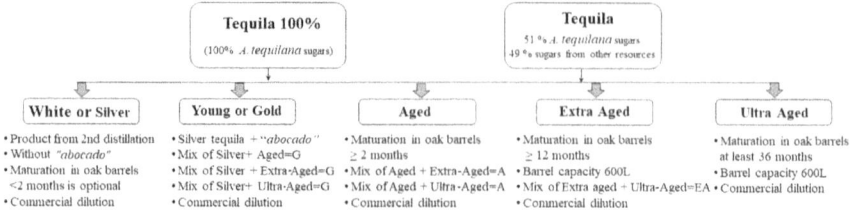

Figure 3.1   Classification of Tequila.

Source: NOM-006-SCFI-2012 (2012).

corn syrup (except other agave species). Both categories of Tequila can be classified into five classes, as shown in Figure 3.1, according to the characteristics obtained after the distillation step (NOM-006-SCFI-2012 2012).

The Tequila's elaboration process begins in the field, with the harvesting and cutting of the agave *piñas*. Once the *piñas* have been transported to the factory, the industrial process takes place in four principal steps. Each step may have some variations impacting directly on the final product quality.

The cooking is done to hydrolyze the polysaccharides in the agave and to soften the tissues to facilitate the grinding and extraction of the cooked juices. The cooking can be done in masonry ovens or pressure autoclaves. Currently, some companies extract raw agave juice by leaching, using diffusers. Then acid-thermal, enzymatic, or combined hydrolysis is performed. This variations in the process can impact the generation of the volatile compounds that give Tequila its characteristic notes.

According to the desired product, the hydrolyzed juices are diluted to achieve the Brix degrees (°Bx, ratio of sugar to water) established by the enterprises before fermentation. Those juices can be enriched with other nutrients. Fermentation can occur spontaneously or be induced by the addition of the commercial *Saccharomyces cerevisiae* yeast or the manufacturer's own inocula prepared in the distillery. The yeast action converts the wort's available sugars into ethanol and other congeners that will define the Tequila's flavor and aroma. Fermented wort distillation can be performed in distillation columns or pot stills; the latter is the more common in the Tequila industry. Traditional distillation using pot stills occurs in two stages. In the first, known as *destrozamiento* (stripping step), a distillate called *ordinario* is obtained with 20–30% alc. vol. Here, the wort solids, part of the water, and the heads (the first distilled fraction) are removed. In the second step, called rectification, the *ordinario* alcohol is concentrated, and the rectified product or heart is obtained. This fraction has an alcohol content 55–60% alc. vol. The head (first), heart (middle), and tail (last) fractions during the second distillation are separated according to the conditions established by the distillery. This double distillate is considered White Tequila and can be commercialized once diluted, filtrated, and bottled. Finally, to the White Tequila can be added authorized additives (caramel color, sugar syrup, oak extract) to obtain Gold Tequila. Additionally, it can be subjected to an aging process into oak or holm barrels for different lengths of time to obtain Aged, Extra-Aged, or Ultra-Aged Tequila.

3.2 VOLATILE COMPOUNDS GENERATION THROUGH THE TEQUILA PROCESS

During Tequila elaboration, each distillery establishes its process conditions, which defines the characteristics that distinguish the different Tequila brands' personalities in the market in its various categories and classes. Each process step plays an important role in the generation, transformation, and depuration of the volatile compounds. Therefore the Tequila quality depends on the control and care throughout each stage of the process.

3.2.1 Raw Material Contribution

Among the great diversity of species of the genus *Agave* recognized in Mexico, only *Agave tequilana* Weber var. Azul can be used in Tequila elaboration. The agave plants are preferably harvested between 6 and 8 years when they reach maturity. Their leaves are immediately cut near the base in a process known as *agave jima* (Cedeño 2003, 1995; Rodríguez Garay *et al.* 2015). After cutting, the agave stems are finally left, often called *piñas* or *agave heads* (Figure 3.2).

The complex reserve carbohydrates in *piñas*, called agave fructans, are commercial interest compounds in the Tequila production process. However, amino acids, pectin, proteins, microelements, and other secondary metabolites are also present: saponins, phenolic compounds, flavonoids, phytosterols, fatty acids, waxes, and terpenic compounds (Sidana, Singh, and Sharma 2016; López-Salazar *et al.* 2019; López-Romero *et al.* 2018). Due to their chemical nature, not all of them contribute to the Tequila aroma in raw juices; however, many will remain or be transformed in subsequent stages.

Among the volatile compounds in the raw stem of *A. tequilana* contribute to the varietal aroma of Tequila; mainly esters, hydrocarbons, furans, organic acids, ketones, alcohols, aldehydes, and terpenes have been found (Prado-Jaramillo *et al.* 2015; Peña-Alvarez *et al.* 2006; Álcazar Valle 2011). The herbal and floral aroma of *piñas* and juices before their hydrolysis are attributed to these last chemical families. Some of the important compounds are 1-octén-3-ol, a-terpineol, linalool, (E)-linalool oxide, (Z)-linalool oxide, g-terpinene, o-cymene, nerolidol, myrcene, limonene, benzene acetaldehyde, (E, Z)-2,4-decadienal, (Z, Z)-2,4-decadienal, (E)-2-heptenal, hexanal, (E)-2-octenal, nonanal, propanal, and vanillin. Details of the raw material's volatile compounds are summarized in Table 3.1.

Figure 3.2   Harvesting and Cutting of *A. tequilana* Weber *var.* Azul to Obtain the Agave Piñas.

TABLE 3.1 MINOR VOLATILE COMPOUNDS IDENTIFIED IN SAMPLES OF DIFFERENT STEPS OF TEQUILA PRODUCTION.

Compounds (IUPAC)	Sample[a]	Ref.[#]	Compounds (IUPAC)	Sample[a]	Ref.[#]
Acetals			**Alcohols**		
1-(1-ethoxyethoxy) pentane	WT, AT,	1, 3	(4-propan-2-ylphenyl) methanol	WT, AT,	3
1-(1-ethoxyethoxy) propane	WT,	1	(S)-3-Ethyl-4-methyl pentanol	WT, AT, EAT	4
1,1,1-triethoxyethane	AT	3	1,2-butanediol	CJ	14
1,1,1-triethoxypropane	AT, EAT	2, 3, 5	1,2-etanediol	CJ	14
1,1,3,3-tetraethoxypropane	WT, AT	3	1,3-butanediol	CJ	14
1,1,3-triethoxypropane	WT	1	1-butanol	CJ, FE, WT, AT, EAT	1, 3, 4, 7 9, 14,
1,1-diethoxy-2-methyl-butane	WT, AT	3, 4	1-decanol	WT, AT	1, 3, 7
1,1-diethoxy-3-methylbutane	WT	1	1-dodecanol	WT, AT, EAT	1, 3, 4, 7
1,1-diethoxy-3-methyl-butane	WT, EAT	4, 5, 7, 8	1-heptanol	WT	1
1,1-diethoxy-butane	WT	1	1-hexadecanol	WT, AT	3, 6, 7, 8, 9
1,1-diethoxy-ethane	FE, WT, AT	1, 3, 6, 7, 8	1-hexanol	RM, CJ, WT, AT	1, 3, 6, 7, 8, 9, 10, 14
1,1-diethoxyhexane	WT, AT	3	1-methoxy-2-butanol	WT	1
1,1-diethoxyisobutane	WT, AT, EAT	2, 3, 5	1-nonanol	WT	8
1,1-diethoxymethane	WT, AT	2, 3	1-octadecanol	WT	1, 8
1,1-diethoxypentane	WT	7, 8	1-octanol	WT	6, 7 9, 10
1,1-diethoxypropane	WT, AT	3, 7, 8	1-pentanol	WT, AT, EAT	3, 4, 7, 9
1,1-dimethoxypropane	WT, AT	3	1-propanol	FE, WT, AT, EAT	1, 3, 5, 7, 9, 11

(Continued)

TABLE 3.1   CONTINUATION

Compounds (IUPAC)	Sample[a]	Ref.[#]	Compounds (IUPAC)	Sample[a]	Ref.[#]
Acetals			**Alcohols**		
1-ethoxy-1-methoxyethane	WT, AT	1, 3, 7	1-tetradecanol	WT, AT	3, 6, 7, 8
1-ethoxy-2-methoxyethane	WT, AT	3,	2,3-butanediol	WT	9
2-(diethoxymethyl)furan	AT	2, 3	2-butanol	WT, AT	3, 8
Acids			2-decanol	WT, AT	3, 8
2,4,5-trioxoimidazolidine	WT, AT	3	2-ethyl-1-butanol	WT	1
2-acetyloxybenzoic acid	CJ	1	2-ethyl-1-hexanoll	WT	10
2-hydroxybutanoic acid	WT	10	2-heptanol	WT, AT	1, 3, 9, 10
2-methylbutanoic acid	AT	2, 3	2-hexanol	WT, AT, EAT	2, 3, 4
2-methylpropanoic acid	RM, FE, WT, AT	1, 3, 7	2-methyl propan-1-ol	FE, WT, AT, EAT	1, 3, 4, 5, 6, 7, 8, 9
3-methylbutanoic acid	WT, AT, EAT	1, 3, 4, 6, 7	2-methyl-3-hexan-1-ol	WT	8
3-phenyl-2-propanoic	CJ	14	2-methyl-3-penten-1-ol	WT, AT	3
4-hydroxybenzeneacetic acid	FE, WT	1, 8	2-methylbut-2-en-1-ol	WT, AT	3, 7, 8
5-(hydroxymethyl)furan-2-carboxylic acid	CJ	1	2-nonanol	WT, AT	3, 8, 1,
benzene acetic acid	CJ	14	2-octanol	WT	6, 8
benzoic acid	CJ	14	2-pentanol	RM, CJ, FE, WT, AT	1, 3
benzoic acid, p-tert-butyl-	WT, AT	2, 3	2-penten-1-ol	WT	9
butanoic acid	CJ, WT, AT	2, 3, 6, 10, 14	2-phenylethanol	RM, CJ, FE, WT, AT, EAT	1, 3, 4, 5
cis-, cis-9,12-octadecadienoic acid	RM, FE	1	3-(1-methylbutoxy)-2-butanol	WT	1,

	[Source]	[Ref]		[Source]	[Ref]
decanoic acid	RM, CJ, WT, AT, EAT	1, 3, 4, 5, 6, 7, 8, 9, 14	3,3-diethoxy propanol	WT	9
dodecanoic acid	CJ, FE, WT, AT, AT, EAT	1, 3, 4, 6, 7, 8, 9, 14	3,4-dimethylpentan-1-ol	WT, AT	2, 3, 10
ethanoic acid	RM, CJ, WT, AT, EAT	1, 3, 4, 7, 8, 9, 10, 14	3-ethoxy-1-propanol	WT, AT	3
furan-2-carboxylic acid	CJ	14	3-ethyl hexanol	WT	9
heptadecanoic acid	RM, CJ	1, 14	3-ethyl-4-methylpentanol	RM	1
heptanoic acid	CJ, WT	1, 6, 14	3-furan methanol	WT	9

[a] RM: raw material; CJ: cooking juice; FE: fermented wort; WT: White Tequila; AT: Aged Tequila; EAT: Extra-Aged Tequila; UAT: Ultra-Aged Tequila. # [1] Prado-Jaramillo et al. (2015); [2] Martín-del-Campo, López-Ramírez, and Estarrón-Espinosa (2019b); [3] Martín-del-Campo, López-Ramírez, and Estarrón-Espinosa (2019a); [4] Aguilar-Méndez et al. (2017); [5] González-Robles and Cook (2016); [6] Estarrón (1997); [7] Benn and Peppard (1996); [8] CIATEJ (1999); [9] Lopez (1999); [10] Lopez and Dufour (1999); [11] Vallejo-Cordoba, Gonzalez-Cordova, and Estrada-Montoya (2004); [12] Rodríguez, Wrobel, Wrobel, and Wrobel (2005); [13] Magana et al. (2015); [14] Mancilla-Margalli and López (2002).

TABLE 3.1 CONTINUATION

Compounds (IUPAC)	Sample[a]	Ref.[#]	Compounds (IUPAC)	Sample[a]	Ref.[#]
Acetals			**Alcohols**		
hexadecanoic acid	RM, CJ, FE, WT, AT	1, 3, 6, 7, 8, 9, 14	3-methyl-1-pentanol	WT, AT	1, 3
hexanoic acid	RM, CJ, WT, AT, EAT	1, 3, 4, 7, 8, 9, 14	3-methyl-3-buten-1-ol	WT, AT	3, 8
nonanoic acid	RM, CJ	1, 14	3-methylbutan-1-ol	RM, CJ FE, WT, AT, EAT	1, 3, 4, 6, 7, 8, 9, 10, 11, 14
octadec-9-enoic acid	CJ	14	3-methylbutan-2-ol	WT	6
octadecanoic acid	CJ	14	3-octanol	WT, AT	3, 6, 7, 10
octanoic acid	RM, CJ, FE, WT, AT, EAT	1, 3, 4, 5, 6, 7, 9, 14	3-pentanol	RM, FE	1
parabanic acid	WT	1	3-penten-1-ol	WT, AT	3, 8
pentadecanoic acid	RM, CJ	1, 14	3-penten-2-ol	WT, AT, EAT	4,
pentanoic acid	CJ, AT	2, 3, 7, 9, 14	3-phenylpropan-1-ol	FE, WT, AT	1, 2, 3, 6
phenylacetic acid	WT	6	4-ethybenziy alcohol		
propanoic acid	RM, CJ FE, WT, AT	1, 3, 6, 9, 10, 14	4-heptanol	WT	1, 10
p-tertbutyl benzoic acid	WT	8	4-hexen-1-ol	WT	10
tetradecanoic acid	RM, CJ, WT, AT	1, 3, 6, 7, 8, 9, 14	4-methylheptan-1-ol	WT	11
Furans			4-methylpentan-1-ol	WT, AT	1, 3, 10
(5-formylfuran-2-yl) methyl acetate	AT	2, 3	4-penten-1-ol	WT	1, 10

Compound	Source	Ref.
(E)-3-(5-methylfuran-2-yl)prop-2-enal	WT	1
1-(2-furanyl) ethanone	CJ, FE, WT, AT, EAT	1, 2, 3, 5
1-(2-furanyl) propanone	WT	8
2-(hydroxymethyl) furan	WT	7, 9
2,5-dimetil-2-(2-tetrahidrofuril) tetrahidrofurano	FE	1
2-acetyl furan	WT	6, 8, 9, 10
2-acetyl-5-methyl furan	WT	9
2-butylfuran	WT, AT	2, 3
2-furan methanol	CJ, FE, WT	1, 14
2-methyldihydrofuran-3-one	WT	9
2-methyltetrahydrofuran-3-one	WT, AT	3, 6
5-(hydroxymethyl) furan-2-carboxaldehyde	CJ, FE, WT, AT	1, 2, 3, 13, 14
5-acetoxymethyl-2-furaldehyde	CJ	1
5-eteniltetrahydro-2-furan	WT	9

Compound	Source	Ref.
4-penten-2-ol	WT	1
5-(hydroxymethyl)-furfuryl alcohol	WT	9
6-methyl heptanol	CJ, WT	8, 14
benzil alcohol	RM, CJ, AT	1, 2, 3
cyclohex-2-en-1-ol	RM	1
ethan-1,2-diol	CJ, WT, AT	1, 2, 6, 7, 14
methanol	WT, AT	3
oct-1-en-3-ol	CJ, WT	
oct-2-en-1-ol	WT, AT	
pent-4-en-1-ol	CJ, WT	6, 7, 8, 9, 10, 14
phenyl ethanol	RM, CJ, FE, WT, AT	1, 3, 14
phenyl methanol	RM, CJ	1, 14
tetramethyl-2-hexadecen-1-ol		
Esters		

(Continued)

TABLE 3.1   CONTINUATION

Compounds (IUPAC)	Sample[a]	Ref.[#]	Compounds (IUPAC)	Sample[a]	Ref.[#]
5-ethenyltetrahydro-R,R-5-trimethyl-2-furanmethanol	CJ	14	(4-ethylphenyl) acetate	FE, WT	1
5-methylfuran-2-carboxaldehyde	CJ, WT, AT, EAT	1, 3, 4, 5, 6, 7, 8, 9, 12, 14	(Z,Z)-9,12-octadecadienoic acid 2-chloroethyl ester	CJ	1
furan-2-carboxaldehyde	CJ, WT, AT, EAT	1, 3, 4, 5, 6, 7, 8, 9, 12, 13	1,4-benzenedicarboxylic acid, dibutyl ester	WT	8
furan-2-yl methanol	WT, AT, EAT	2, 3, 4	1-phenylpropyl acetate	WT, AT	2, 3
furan-2-ylmethyl formate	WT, AT	2, 3	2-hydroxyethyl propionate	FE, WT	1
tetrahydro-2-methylfuran	CJ	14	2-methylpropyl acetate	WT, AT	1, 3
(E)-2-heptenal	RM, CJ	1	2-methylpropyl hexanoate	WT	1
(E)-2-octenal	RM	1	2-phenylethyl 3-methylbutanoate	WT	1, 6
(E,E)-2,4-hexadienal	WT	1	2-phenylethyl formate	WT	10
(E,Z)-2,4-decadienal	RM, CJ	1	2-phenylethyl hexanoate	WT	1

[a] RM: raw material; CJ: cooking juice; FE: fermented wort; WT: White Tequila; AT: Aged Tequila; EAT: Extra-Aged Tequila; UAT: Ultra-Aged Tequila. [#] 1 Prado-Jaramillo et al. (2015); [2] Martín-del-Campo, López-Ramírez, and Estarrón-Espinosa (2019b); [3] Martín-del-Campo, López-Ramírez, and Estarrón-Espinosa (2019a); [4] Aguilar-Méndez et al. (2017); [5] González-Robles and Cook (2016); [6] Estarrón (1997); [7] Benn and Peppard (1996); [8] CIATEJ (1999); [9] Lopez (1999); [10] Lopez and Dufour (1999); [11] Vallejo-Cordoba, Gonzalez-Cordova, and Estrada-Montoya (2004); [12] Rodríguez, Wrobel, and Wrobel (2005); [13] Magana et al. (2015); [14] Mancilla-Margalli and López (2002).

Volatile Compounds Formation in Beverages   43

Aldehydes		
(Z,Z)-2,4-decadienal	RM	1
1-mercapto-2-heptadecanone	FE, WT	1
1-p-menthen-9-al	WT	1
2,3-dihydro-2-methyl-4(H)-pyran-4-one	CJ	14
2,3-dihydro-4(H)-pyran-4-one	CJ	14
2,3-dihydroxy-3,5-dihydro-6-methyl-4(H)-pyran-4-one	CJ	14
2,6,6-trimethylcyclohexa-1,3-diene-1-carbaldehyde	WT, AT	3
2,6,6-trimethylcyclohexene-1-carbaldehyde	AT	2, 3
2-methyl-4-hydroxybenzaldehyde	WT	1
2-methyl-6-methylen-2,7-octadien-1-ol	WT	1
2-pentadecanone	WT	1
2-phenylacetaldehyde	AT	2, 3

Esters		
2-methylpropyl dodecanoate	WT	1, 9
2-phenylethyl propionate	WT, AT	3
3,7,11-trimethyldodeca-2,6,10-trienyl acetate	WT	1
3,7-dimethyl-1,6-octadiene-3,4-diol	CJ	14
3,7-dimethylocta-1,6-dien-3-yl propanoate	CJ, WT	9, 14
3-methyl ethyl decanoate	WT	11
3-methyl-1-propanyl acetate	WT, AT, EAT	4
3-methylbutyl acetate	RM, CJ, FE, WT, AT, EAT	1, 3, 4, 5, 6, 7
3-methylbutyl decanoate	WT, AT, EAT	1, 3, 4, 8, 11
3-methylbutyl dodecanoate	WT	1
3-methylbutyl octanoate	WT, AT	3, 8
6-methy heptyl-2-propanoate	WT	9
acetic acid 2-phenylethyl ester	WT, AT EAT	3, 4, 5

(*Continued*)

TABLE 3.1   CONTINUATION

Compounds (IUPAC)	Sample[a]	Ref.[#]	Compounds (IUPAC)	Sample[a]	Ref.[#]
2-phenylacetaldehyde	CJ	1	benzyl acetate	WT	8
3-(ciclohex-2'-en-il)-propionaldehyde	WT	1	benzyl benzoate	RM, CJ, WT, AT	1, 3, 8
3,5-dihydroxy-6-methyl-2,3-dihydropyran-4-one	CJ, FE	1	bis(2-ethylhexyl) benzene-1,2-dicarboxylate	WT, AT	3
3-ethoxypropanal	WT, AT	3, 8, 10	Butanedioic acid, diethyl ester	FE, WT, AT, EAT	1, 3, 4, 5 6, 8, 10
3-hydroxy-2-methyl-4H-pyran-4-one	CJ	14	butyl decanoate	WT	11
3-hydroxy-2-methylbenzaldehyde	WT	1	butyl hexadecanoate	CJ, FE	1
3-methylbutanal	WT, AT	3	butyl isobutyl phthalate	RM, WT	1
4-ethylbenzaldehyde	WT	9	dibutyl 1,2-benzenedicarboxylate	FE	1
4-hydroxy-3,5-dimethoxybenzaldehyde	CJ, AT	2, 3, 14	dibutyl benzene-1,4-dicarboxylate	WT, AT	3
4-hydroxy-3-methoxybenzaldehyde (vanillin)	CJ, WT, AT, EAT	1, 3, 4, 5, 7, 9, 10, 14,	dibutyl phthalate	RM, CJ, WT	1, 6
4-hydroxybenzaldehyde	CJ	14	dioctyl hexanedioate	FE, WT	1, 6
4-propan-2-ylcyclohexene-1-carbaldehyde	AT	2, 3	dioctyl phthalate	WT	8
acetaldehyde	WT	12	ethyl (2S)-2-hydroxypropanoate	WT, AT, EAT	3, 4
benzaldehyde	CJ, WT, AT	3, 7, 8, 10, 14	ethyl (9Z,12Z)-octadeca-9,12-dienoate	RM, CJ, FE, WT, AT, EAT	1, 3, 4, 7, 8, 9, 10, 11

formaldehyde	WT	[12]	ethyl (9Z,12Z,15Z)-octadeca-9,12,15-trienoate	RM, FE, WT, AT	[1, 3, 7, 8, 11]
hexanal	RM, CJ, WT, AT	[1, 3, 8, 10]	ethyl (Z)-octadec-9-enoate	RM, FE, WT, AT	[1, 3, 7, 8, 10]
nonanal	RM, CJ,	[1]	ethyl 2-furancarboxylate	WT, AT	[3, 6, 8]
octodecanal	RM, CJ, WT	[1]	ethyl 2-hydroxy-3-methylbutanoate	WT, AT	[3, 6]
phenyl acetaldehyde	CJ	[14]	ethyl 2-hydroxy-3-phenylpropanoate	FE, WT	[1]
propanal	RM	[1]	ethyl 2-hydroxy-butanoate	WT	[9]
tetradecanal	WT	[1]	ethyl 2-hydroxy-propanoate	WT, AT, EAT	[3, 5, 6, 7, 8, 9, 10]
(+)-car-2-en-4-one	WT	[8]	ethyl 2-methyl butanoate	AT	[2, 3, 9]
1-(2-furanyl)ethanone	CJ	[14,]	ethyl 2-oxopropanoate	WT	[1, 7]
			ethyl 4-hydroxybutanoate	FE	[1]

[a] RM: raw material; CJ: cooking juice; FE: fermented wort; WT: White Tequila; AT: Aged Tequila; EAT: Extra-Aged Tequila; UAT: Ultra-Aged Tequila. # [1] Prado-Jaramillo et al. (2015); [2] Martín-del-Campo, López-Ramírez, and Estarrón-Espinosa (2019b); [3] Martín-del-Campo, López-Ramírez, and Estarrón-Espinosa (2019a); [4] Aguilar-Méndez et al. (2017); [5] González-Robles and Cook (2016); [6] Estarrón (1997); [7] Benn and Peppard (1996); [8] CIATEJ (1999); [9] Lopez (1999); [10] Lopez and Dufour (1999); [11] Vallejo-Cordoba, Gonzalez-Cordova, and Estrada-Montoya (2004); [12] Rodríguez, Wrobel, and Wrobel (2005); [13] Magana et al. (2015); [14] Mancilla-Margalli and López (2002).

TABLE 3.1 CONTINUATION

Compounds (IUPAC)	Sample[a]	Ref.#	Compounds (IUPAC)	Sample[a]	Ref.#
Ketones			**Esters**		
(E)-1-(2,6,6-trimethylcyclohexa-1,3-dien-1-yl)but-2-en-1-one	CJ, WT, AT	1, 3, 6, 7, 14	ethyl 2-methyl propanoate	WT, AT	3
(E)-4-(2,6,6-trimethylcyclohex-2-en-1-yl)but-3-en-2-one	WT, AT, EAT	3, 5	ethyl 2-phenyl octanoate	WT	11
1-(4-hydroxy-3-methoxyphenyl)ethanone	RM, CJ	1	ethyl 3-methylbutanoate	WT, AT	3, 8
1-(4-hydroxyphenyl)ethanone	WT	1	ethyl 4-oxopentanoate	WT, AT	3, 6, 8, 10
1-(6-bicyclo[2.2.1]hept-2-enyl)ethanone	WT	1	ethyl 9,9-diethoxynonanoate	WT	1
1-[2-(1-methylethyl)-1H-imidazol-4-yl]-ethanone	WT	1	ethyl 9-decanoate	WT	6, 7, 8
1-furan-2-ylpropan-2-one	CJ	1	ethyl 9-decenoate	WT	1, 6
1-hydroxy-2-propanone	CJ	14	ethyl 9-hexadecenoate	WT	10
1-phenylpropan-2-one	WT	1	ethyl 9-oxo-nonanoate	WT	1
2(5H)-furanone	CJ	14	ethyl acetate	FE, WT, AT	1, 3, 6, 7 8,
2,3-pentanedione	WT	8	ethyl benzoate	WT, EAT	5
2,5-hexanedione	FE,	1	ethyl butanoate	WT, AT	1, 3, 6, 7, 8
			ethyl decanoate	FE, WT, AT, EAT	1, 3, 4, 5, 6, 7, 8, 9, 10, 11
2-buten-1-one	WT	11	ethyl dodecanoate	WT, AT, EAT	1, 4, 5, 6, 8, 9, 10, 11

Compound	Sources	References
2-cyclohexen-1-one	RM, CJ, WT	1
2-ethylcyclobutanone	CJ, WT	1
2-methylcyclopentanone	WT	1, 8, 9
2-methyltetrahydrothiophen-3-one	FE, WT	1
3,3-diethoxybutan-2-one	WT	1
3-ethyl-2-hydroxycyclopent-2-en-1-one	CJ	1
3-hydroxy-2-butanone	RM, CJ, FE, WT, AT, EAT	1, 4, 5, 14
3-hydroxybutan-2-one	WT	8, 9
3-methyl-2(5H)-furanone	CJ	14,
3-methyl-2-butanone	WT	1
3-methyl-2-pentanone	WT	1
3-methylcyclopentan-1-one	FE, WT, AT	1, 3, 6, 7
3-methylcyclopentane-1,2-dione	CJ	14
3-penten-one	WT	6

Compound	Sources	References
ethyl formate	WT, AT	2, 3
ethyl hexadec-9-enoate	FE, WT	1
ethyl hexadecanoate	RM, CJ, FE, WT, AT, EAT	1, 3, 4, 5, 6, 7, 8, 9, 10, 11
ethyl hexanoate	FE, WT, AT, EAT	1, 3, 4, 5, 6, 7, 8, 9, 10
ethyl nonanoate	WT	7, 8
ethyl octadecanoate	FE, WT	1, 9, 11
ethyl octanoate	RM, CJ, FE, WT, AT, EAT	1, 3, 4, 5, 6, 8, 9, 10, 11
ethyl pentadecanoate	FE, WT, AT	1, 3
ethyl pentanoate	WT, AT,	1, 3, 8
ethyl propanoate	WT, AT	3, 6, 7
ethyl tetradecanoate	FE, WT, AT, EAT	1, 3, 4, 5, 7, 8, 9, 10, 11
ethyl-2-hydroxy-4-methyl-pentanoate	WT, AT, EAT	4, 6, 7
furan-2-ylmethyl formate	WT	1
furfuryl acetate	CJ, WT	1, 6

(Continued)

TABLE 3.1 CONTINUATION

Compounds (IUPAC)	Sample[a]	Ref.[#]	Compounds (IUPAC)	Sample[a]	Ref.[#]
5-methyl-3-hexen-2-one	CJ	1	furfuryl formate	WT	6, 8
butane-2,3-dione	WT, AT,	3	isopenthyl hexanoate	WT	1
cyclopentanone	FE, WT, AT, EAT	1, 3, 4, 6, 7	methyl (Z)-octadec-9-enoate	RM	1
dihydro-2(3H)-furanone	FE, WT	1	methyl (Z,Z)-9,12-octadecadienoate	RM, FE	1
dihydro-5-methyl-2(3H)-furanone	CJ	14	methyl 10,13-octadecadienoiate	RM	1
hexan-2-one	WT, AT	3	methyl 2-furoate	CJ, WT	1, 8, 14

[a] RM: raw material; CJ: cooking juice; FE: fermented wort; WT: White Tequila; AT: Aged Tequila; EAT: Extra-Aged Tequila; UAT: Ultra-Aged Tequila. [#] [1] Prado-Jaramillo et al. (2015); [2] Martín-del-Campo, López-Ramírez, and Estarrón-Espinosa (2019b); [3] Martín-del-Campo, López-Ramírez, and Estarrón-Espinosa (2019a); [4] Aguilar-Méndez et al. (2017); [5] González-Robles and Cook (2016); [6] Estarrón (1997); [7] Benn and Peppard (1996); [8] CIATEJ (1999); [9] Lopez (1999); [10] Lopez and Dufour (1999); [11] Vallejo-Cordoba, Gonzalez-Cordova, and Estrada-Montoya (2004); [12] Rodríguez, Wrobel, and Wrobel (2005); [13] Magana et al. (2015); [14] Mancilla-Margalli and López (2002).

TABLE 3.1 CONTINUATION

Compounds (IUPAC)	Sample[a]	Ref.[#]	Compounds (IUPAC)	Sample[a]	Ref.[#]
Ketones			**Esters**		
dihydro-2-methyl-3(2H)-furanone	CJ, FE, WT, AT, EAT	1, 3, 14	methyl (Z,Z,Z)-9,12,15-octadecatrienoate	RM	1
hexan-3-one	WT	11	methyl 2-hydroxy-3-phenyl propanoate	WT	6
pentan-2-one	RM, CJ, FE, WT, AT	1, 3	methyl 2-hydroxybenzoate	WT, AT	1, 3, 6, 8
pentane-2,3-dione	CJ, WT	1	methyl 3-furoate	FE, WT	1
Divers			methyl 4-methyl benzoate	WT	6
(1R,2S,6S,7S,8S)-8-isopropyl-1,3-dimethyltricyclo[4.4.0.02,7]dec-3-ene	WT	1	methyl decanoate	WT	11
(3-hydroxyphenyl) acetate	WT, AT	3	methyl dodecanoate	WT	11
(6E,10E,14E,18E)-2,6,10,15,19,23-hexamethyltetracosa-2,6,10,14,18,22-hexaene	WT	14	methyl hexadecadienoate	WT	1
(7s,10s,5e)-2,6,10-trimetil-7,10-epoxi-2,5,11-dodecatrieno	WT	1	methyl hexacecanoate	RM, CJ, WT, AT	1, 2, 3
[2S-[2alpha,5beta(E)]]-5-(1,5-dimethyl-1,4-hexadienyl)-2-ethenyltetrahydro-2-methylfuran	WT, AT	2, 3	methyl hexanoate	WT	6
1-(2,6-dihydroxy-4-methoxyphenyl)ethanone	WT, AT	2, 3	methyl nonanoate	AT	2, 3
1,1,6-trimethyl-2H-naphthalene	WT	1	methyl octadecanoate	WT,	1

(Continued)

TABLE 3.1 CONTINUATION

Compounds (IUPAC)	Sample[a]	Ref.[#]	Compounds (IUPAC)	Sample[a]	Ref.[#]
1,1,6-trimethyl-2H-naphthalene	WT, AT	3	pentyl 2-hydroxypropanoate	WT, AT	2, 3
1,2,3,4,4a,5,6,7,8,8a-decahydronaphthalene	AT	2, 3	phenethyl octadecanoate	WT	10,
1,2,3,5-tetramethylbenzene	WT	8	phenethyl octanoate	WT	1, 9
1,2-dichlorobenzene	RM, WT	1	phenylethyl acetate	WT	1, 7, 8, 9, 10
1,4-dichlorobenzene	RM, CJ, WT	1	phenylpropyl acetate	WT	7, 8
1,6-dimethylnaphthalene	WT, AT,	3	prop-2-enyl 2-phenylacetate	WT, AT	2, 3
1.beta.-Cadin-4-en-10-ol	WT	8	propyl decanoate	WT	11
1H-pyrrole-2-carboxaldehyde	CJ	14	**Phenols**		
1-methyl pyrazol	WT	9	(1,1-dimethylethyl)-4-methoxyphenol	WT	1
1-propyldisulfanyl propane	RM	1	1,3-bencenediol monoacetate	WT	6
2,10,10-trimethyl-6-methylidene-1-oxaspiro[4.5]dec-7-ene	WT	6	1,4-bencenediol	WT	8
2,2,6-trimethyl-6-vinyltetrahydro-2H-pyran	WT, AT	3	1,4-bencenediol monoacetate	WT	6
2,3-dihydrothiophene	CJ	14	1,4-bencenediol-5-methyl	WT	6
2,3-dimethylthiophene	CJ	14	1-ethoxy-4-ethylbenzene	WT	6
2,4,5-trimethylthiazole	CJ	14	2,3,6-trimethylphenol	WT, AT	3
2-[2-(methoxy)ethyl]-4(5)-methylimidazole	WT,	1	2,5-di-tert-butylphenol	WT	1, 10

Compound	Source	Ref.	Compound	Source	Ref.
2-cyclopropylthiophene	CJ	[14]	2,6-diethylphenol	WT	[1]
2-ethoxy thiazol	WT	[10]	2,6-dimethoxy-4-prop-2-enylphenol	WT, AT	[3]
2-formyl-1-methylpyrrole	WT,	[1]	2,6-dimethoxyphenol	AT, EAT	[2, 3, 5]
2-methoxy-5-(2-methylpropyl)-	CJ	[14]	2,6-ditert-butyl-4-methylphenol	FE, WT, AT	[1, 3, 8]
2-propan-2-yldisulfanylpropane	CJ	[1]	2-[(3R,5S,8R)-3,8-dimethyl-1,2,3,4,5,6,7,8-octahydroazulen-5-yl]propan-2-ol	WT	[1]
2-pyrrolidinone	CJ	[14]	2-isopropyl-5-methylphenol	CJ, FE, WT, AT	[1, 3, 8, 9, 10]
3-(methylthio)propanol	FE,	[1]	2-methoxy-4-methylphenol	WT, AT	[3, 6, 8]
3,4,5-trimethyl pyrazol	WT	[9]	2-methoxy-4-prop-2-enylphenol	CJ, WT, AT	[1, 3]
3,4-diethyl-2-methylpyrrole	CJ	[14]	2-methoxy-4-propylphenol	WT, AT, EAT	[2, 3, 5]

[a] RM: raw material; CJ: cooking juice; FE: fermented wort; WT: White Tequila; AT: Aged Tequila; EAT: Extra-Aged Tequila; UAT: Ultra-Aged Tequila. # [1] Prado-Jaramillo et al. (2015); [2] Martín-del-Campo, López-Ramírez, and Estarrón-Espinosa (2019b); [3] Martín-del-Campo, López-Ramírez, and Estarrón-Espinosa (2019a); [4] Aguilar-Méndez et al. (2017); [5] González-Robles and Cook (2016); [6] Estarrón (1997); [7] Benn and Peppard (1996); [8] CIATEJ (1999); [9] Lopez (1999); [10] Lopez and Dufour (1999); [11] Vallejo-Cordoba, Gonzalez-Cordova, and Estrada-Montoya (2004); [12] Rodríguez, Wrobel, and Wrobel (2005); [13] Magana et al. (2015); [14] Mancilla-Margalli and López (2002).

TABLE 3.1 CONTINUATION

Compounds (IUPAC)	Sample[a]	Ref.[#]	Compounds (IUPAC)	Sample[a]	Ref.[#]
Divers			**Phenols**		
3,4'-difluoro-4-metoxibifenilo	WT	1	2-methoxyphenol	FE, WT, AT	1, 2, 3, 6, 7
3,4-Dihydro-2H-pyran	WT	9	2-methyl-5-propan-2-ylphenol	WT, AT	3
3,7-dimethyl-1,3,7-octatriene	RM, WT	1, 8	2-methylphenol	WT	1, 7
3,7-dimethyl-dibenzothiophene	WT, AT, EAT	4	2-tert-butylphenol	WT, AT	3
3-furanacetic acid, 4-hexyl-2,5-dihydro-2,5-dioxo	AT,	2, 3	3,4-diethyl phenol	WT	6, 10
3-methyl-dibenzothiophene	WT, AT, EAT	4	3,4-dimethyl phenol	WT	10
3-methylpent-2-ene	WT, AT	3	4-(ethoxymethyl)phenol	WT,	1
3-methylstyrene	WT	1	4-allyl-2-methoxyphenol	RM, CJ, WT	1, 6, 7, 8
4,4-dimethyl-8-methylene-1-oxaspiro[2.5]octane	AT	2, 3	4-ethyl-2-methoxyphenol	WT, AT, EAT	3, 5, 6, 8
4,5-dimethyl-2-propyl-thiazole	CJ	14	4-methylphenol	WT	7, 10
4,7-dimethyl-1-propan-2-yl-1,2-dihydronaphthalene	WT	1	4-prop-2-enylphenol	WT, AT	2, 3
4-cyclopentene-1,3-dione	CJ,	1	5-isopropyl-2-methylphenol	CJ, FE, WT	1, 8, 10
4-ethyl-2-methylpyrrole	CJ	14	**Terpenes**		
4-methyl-dibenzothiophene	WT, AT, EAT	4	(1R,4S,4aR,8aS)-4-isopropyl-1,6-dimethyl-1,2,3,4,4a,7,8,8a-octahydro-1-naphthalene	WT, AT	2, 3
4-methyl-naphtho-(1,2)-thiophene	WT, AT, EAT	4	(1S,4aS,8aR)-1-isopropyl-4,7-dimethyl-1,2,4a,5,6,8a-hexahydronaphthalene	WT	1, 8

Compound	Source	Ref	Compound	Source	Ref
9-hidroxipirimido[1,6-a]pirimidin-4-ona	WT	1	(1S,8aR)-4,7-dimethyl-1-propan-2-yl-1,2,3,5,6,8a-hexahydronaphthalene	RM, WT, AT	1, 2, 3, 8
benzene-1,4-diol	WT, AT	2, 3	(2E)-3,7-dimethylocta-2,6-dien-1-ol	WT, AT	3
bis(2-ethylhexyl)phthalate (BEHP)	WT	1	(2E,6E)-3,7,11-trimethyldodeca-2,6,10-trien-1-ol	FE, WT, AT	1, 3, 6, 8
di(phenyl)methanone	RM,	1	(2Z)-3,7-dimethylocta-2,6-dien-1-ol	WT, AT	3
dihidro-2-metil-3(2H)-tiofeno	FE, WT	1	(3R)-3,7-dimethyloct-6-en-1-ol	WT	1, 6
disulfide, 1-methylethyl propyl	CJ	1	(4-propan-2-ylphenyl)methanol	FE, WT	1, 6, 8
methyldisulfanylmethane	WT	1	(5E)-2,6-dimethylocta-5,7-dien-2-ol	WT	1
N,N-dimethyl-octanamide	WT, AT, EAT	4	(S)-6-ethenyl-6-methyl-1-(1-methylethyl)-3-(1-methylethylidene)cyclohexene	WT	6
N-acetyl-L-alanine	CJ	14	1-methyl-4-prop-1-en-2-ylcyclohexan-1-ol	WT	8
vulgarol B	WT	1	1-methyl-4-propan-2-yl-7-oxabicyclo[2.2.1]heptane	WT, AT	2, 3
Hydrocarbons					
(+)-cycloisosativene	RM, WT, AT	1	1-methyl-4-propan-2-ylbenzene	WT, AT	3
			1-methyl-4-propan-2-ylcyclohex-3-en-1-ol	AT	2, 3
1,2,3-tetramethyl benzene	WT	6	2-(4-methyl-1-cyclohex-3-enyl)propan-2-ol	RM, CJ, FE, WT, AT, EAT	1, 3, 4, 5, 6, 8
1,3-hexadiene,3-ethyl-2-methyl	RM,	1	2-(4-methylcyclohex-3-en-1-yl)propan-2-ol	WT, AT	3

(Continued)

Compounds (IUPAC)	Sample[a]	Ref.[#]	Compounds (IUPAC)	Sample[a]	Ref.[#]
1,6-dimethylpyridineethi-2-one	WI	9	2-(5-ethenyl-5-methyloxolan-2-yl)propan-2-ol	FE, WT, AT	1, 3
1-hexadecylene	WT	1	2,6,6-trimethylcyclohex-2-ene-1-carbaldehyde	WT, AT	1, 3
1-methyl-1-ethylcyclopentane	RM	1	2,6,6-trimethylcyclohexene-1-carbaldehyde	WT	7, 8
1-methyl-2-(1-methylethyl)benzene	WT	1	2-[(2R,5R)-5-methyl-5-vinyltetrahydrofuran-2-yl]propan-2-ol	CJ, FE, WT	1, 8, 9

[a] RM: raw material; CJ: cooking juice; FE: fermented wort; WT: White Tequila; AT: Aged Tequila; EAT: Extra-Aged Tequila; UAT: Ultra-Aged Tequila. [#] 1 Prado-Jaramillo *et al.* (2015); [2] Martín-del-Campo, López-Ramírez, and Estarrón-Espinosa (2019b); [3] Martín-del-Campo, López-Ramírez, and Estarrón-Espinosa (2019a); [4] Aguilar-Méndez *et al.* (2017); [5] González-Robles and Cook (2016); [6] Estarrón (1997); [7] Benn and Peppard (1996); [8] CIATEJ (1999); [9] Lopez (1999); [10] Lopez and Dufour (1999); [11] Vallejo-Cordoba, Gonzalez-Cordova, and Estrada-Montoya (2004); [12] Rodríguez, Wrobel, and Wrobel (2005); [13] Magana *et al.* (2015); [14] Mancilla-Margalli and López (2002).

TABLE 3.1 CONTINUATION

Compounds (IUPAC)	Sample[a]	Ref.[#]
Hydrocarbons		
2,6,10-trimethyl tetradecane	WT, AT, EAT	4
2-hydroxypyridine	CJ	14
2-methylnonadecane	WT	1
3,4-dimethylpyridine	WT	9
3-methy-2-pentene	WT	8, 9
3-methyl-octadecane	WT, AT, EAT	4
3-methylpentane	WT	1
7-methyl-1-octene	WT	1
9-eicosane	WT	9
9-hidroxipirimido[1,6-a]pirimidin-4-ona	WT	1
cyclohexadecane	FE	1
cyclotetradecane	WT	9,
docosane	WT, AT, EAT	4,
eicosane	WT, AT, EAT	4
ethenylbenzene	WT	1

Compounds (IUPAC)	Sample[a]	Ref.[#]
Phenols		
2-[(2R,5R)-5-methyl-5-vinyltetrahydrofuran-2-yl]propan-2-ol	WT, AT, EAT	3, 5
2-[(2R,5S)-5-methyl-5-vinyltetrahydrofuran-2-yl]propan-2-ol	CJ, FE, WT	1, 6, 8
2-[(2S,5S)-5-methyl-5-(4-methylcyclohex-3-en-1-yl]oxolan-2-yl]propan-2-ol;(3S,6S)-2,2,6-trimethyl-6-[(1S)-4-methylcyclohex-3-en-1-yl]oxan-3-ol	WT, AT	3
2-[5-methyl-5-(4-methyl-3-cyclohexen-1-yl)-tetrahydro-2-furanyl]propan-2-ol	WT	1
2-[5-methyl-5-(4-methylcyclohex-3-en-1-yl)oxolan-2-yl]propan-2-ol	WT,	1, 8
3,3-dimethyl-2-methylidenebicyclo[2.2.1]heptane	WT, AT	3
3,7,11,15-tetramethylhexadec-1-en-3-ol	WT	1
3,7,11-trimethyl-1,6,10-dodecatrien-3-ol	RM, CJ, WT, AT, EAT	1, 3, 4, 6, 9
3,7,11-trimethyldodeca-6,10-dien-1-ol	WT	6, 8
3,7,7-trimethylbicyclo[4.1.0]hept-3-ene	WT,	1,
3,7-dimethyl-2E,6-octadien-1-ol	CJ	14
3,7-dimethyl-2Z,6-octadien-1-ol	CJ, WT	6, 14

(Continued)

TABLE 3.1   CONTINUATION

Compounds (IUPAC)	Sample[a]	Ref.[#]	Compounds (IUPAC)	Sample[a]	Ref.[#]
heneicosane	CJ, WT, AT, EAT	1,4	3,7-dimethyloct-6-en-1-ol	WT, AT, EAT	3,5
heptadecane	RM, CJ, WT, AT, EAT	1,4,	3,7-dimethylocta-1,6-dien-3-ol	RM, CJ, FE, WT, AT, EAT	1,3,4,5,6, 8,9,14,
hexadecane	WT, AT, EAT	4	3-methyl-6-propan-2-ylidenecyclohexene	RM	1
hexadecane	CJ	1	4-methyl-1-(1-methylethyl)-1,4-cyclohexadiene	RM	1
methylbenzene	WT	6	4-methyl-1-propan-2-ylcyclohex-3-en-1-ol	WT, AT,	1,3,6,8
n-eicosane	RM, CJ	1	farnesol A isomer	WT	1
nonadecane	RM	1	lactones		
octacosane	WT, AT, EAT	4	5-methyl-3H-furan-2-one	WT, AT	1,3,6
octadecane	WT, AT, EAT	4,	dihydro-5-pentyl-2(3H)-furanone	RM	1,8,9
pyridine	CJ, WT	1,14,	dihydrofuran-2(3H)-one	CJ	14
tetracosane	CJ, WT, AT, EAT	1,4	5-propyloxolan-2-one	AT	2,3
tetradec-1-ene	WT	1	5-butyl-4-methyloxolan-2-one	AT	2,3
tetradecane	WT	1,	5-methyl decahydro 3(2H)furanone	WT, AT	2,3
triacontane	WT, AT, EAT	4	1-(5-methylfuran-2-yl)ethanone	WT, AT	3
			(4R,5S)-5-butyl-4-methyldihydrofuran-2(3H)-one	EAT	5
			(4S,5S)-5-butyl-4-methyldihydrofuran-2(3H)-one	EAT	5

[a] RM: raw material; CJ: cooking juice; FE: fermented wort; WT: White Tequila; AT: Aged Tequila; EAT: Extra-Aged Tequila; UAT: Ultra-Aged Tequila. [#] [1] Prado-Jaramillo et al. (2015); [2] Martín-del-Campo, López-Ramírez, and Estarrón-Espinosa (2019b); [3] Martín-del-Campo, López-Ramírez, and Estarrón-Espinosa (2019a); [4] Aguilar-Méndez et al. (2017); [5] González-Robles and Cook (2016); [6] Estarrón (1997); [7] Benn and Peppard (1996); [8] CIATEJ (1999); [9] Lopez (1999); [10] Lopez and Dufour (1999); [11] Vallejo-Cordoba, Gonzalez-Cordova, and Estrada-Montoya (2004); [12] Rodríguez, Wrobel, and Wrobel (2005); [13] Magana et al. (2015); [14] Mancilla-Margalli and López (2002).

3.2.2 Agave Cooking

In the traditional process, the cooking of agave *piñas* aims to hydrolyze fructans from *Agave tequilana* Weber var. Azul to obtain fermentable sugars, mainly fructose and glucose, assimilable by yeast to produce ethanol (Cedeño 1995). Simultaneously, the generation of a great variety of volatile compounds that significantly contribute to Tequila's aroma and flavor is favored by the chemical reactions occurring during this thermal process. The generation of compounds, whether by oxidation-dehydration, sugar caramelization, or Maillard reactions, is favored by various factors. The agave age, type of cooking/hydrolysis, pH, time, and temperatures all affect the development of volatiles at this stage (Mancilla-Margalli and López 2002; Waleckx *et al.* 2008; Cedeño 1995; Tellez Mora 2000). The research focused on the evolution of volatile compounds during the cooking of agave *piñas* has shown that furans such as 5-hydroxymethylfurfural, furfural, 2-acetylfuran, 5-methyl furfural, 2-methyltetrahydrofuran-3-one, and tetrahydro-2-methylfuran are, in quantitative terms, the main compounds generated (Figure 3.3a) (Cosío Ramirez *et al.* 2002; Estarrón Espinosa, Martín del Campo Barba, and Pinal Suazo 2001; Waleckx *et al.* 2008; Mancilla-Margalli and López 2002; Pinal 2001).

Additionally, other furans and compounds of different chemical families, including organic acids, alcohols, aldehydes, ketones, esters, phenols, terpenes, sulfur compounds, and pyrans, contribute the special and pleasant aroma of cooked *piñas* (Figures 3.3b and 3.3c). About 219 minor volatile compounds present in agave juices after heat treatment have been reported (Mancilla-Margalli and López 2002; Prado-Jaramillo *et al.* 2015; Estarrón Espinosa, Martín del Campo Barba, and Pinal Suazo 2001); some are described in Table 3.1. On the other hand, the generation of methanol, an important major compound regulated by the Official Standard, takes place mainly at this stage. Methanol is formed during cooking due to pectins' demethoxylation of the *A. tequilana* Weber var. Azul in the presence of water. Its production is favored by high cooking temperatures, low pH, use of young agaves, *jima* height, long waiting times between the agave harvest and its entry into the process (Cedeño 1995, 2003; Pérez Martínez *et al.* 2015). Due to its toxicity, 300 mg/100 mL a.a. of methanol is allowed in Tequila, so the generation of this compound in the cooking stage is critical. The elimination of steam through exhaust valves during cooking and reasonable control of the tail cuts in the distillation can make it possible to maintain the permitted methanol concentration concentrations (Solís-García *et al.* 2017).

(a)

5-Hydroxymethylfurfural   Furfural
2-Acetylfuran   2-Methyltetrahydrofuran-3-one
5-Methylfurfural

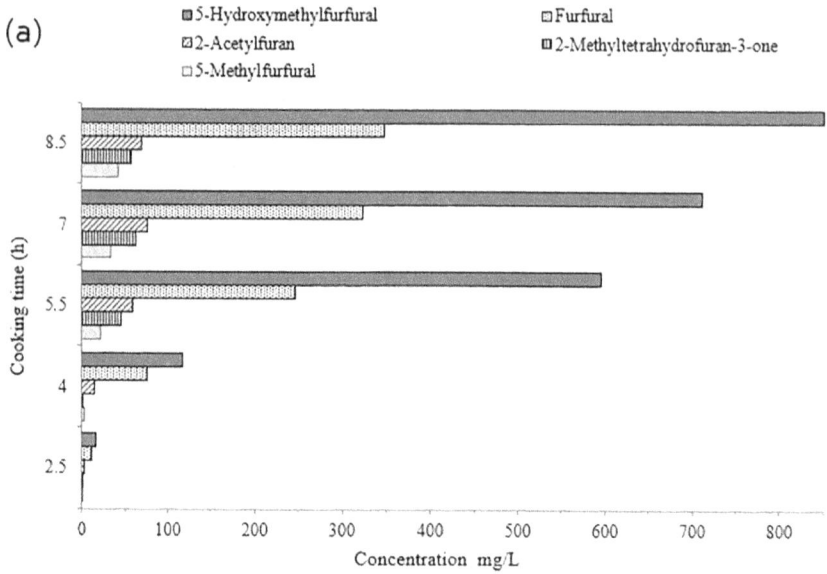

(b)

Ethyl lactate   3-Octanol   x — Ethyl furoate
Phnetyl acetate   Phenetyl alcohol   Vanillin

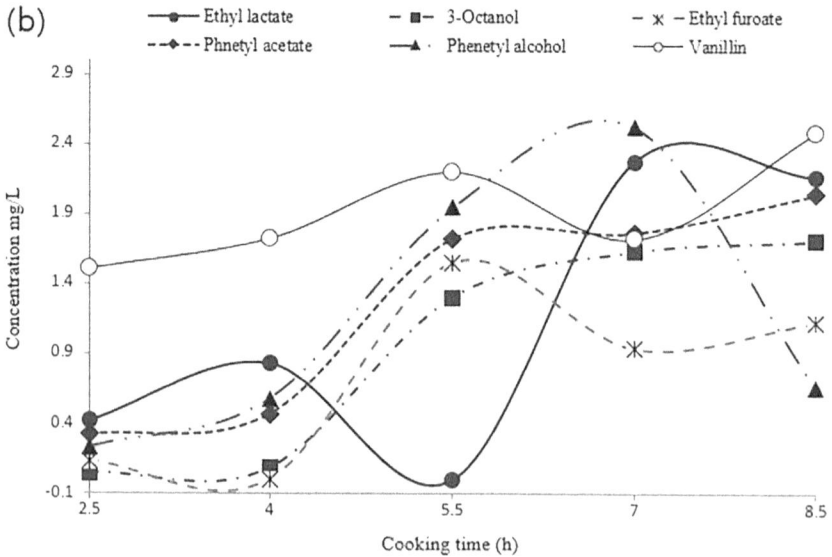

Figure 3.3   Changes in the Generation of Volatile Compounds during the Laboratory-Scale Cooking Stage.

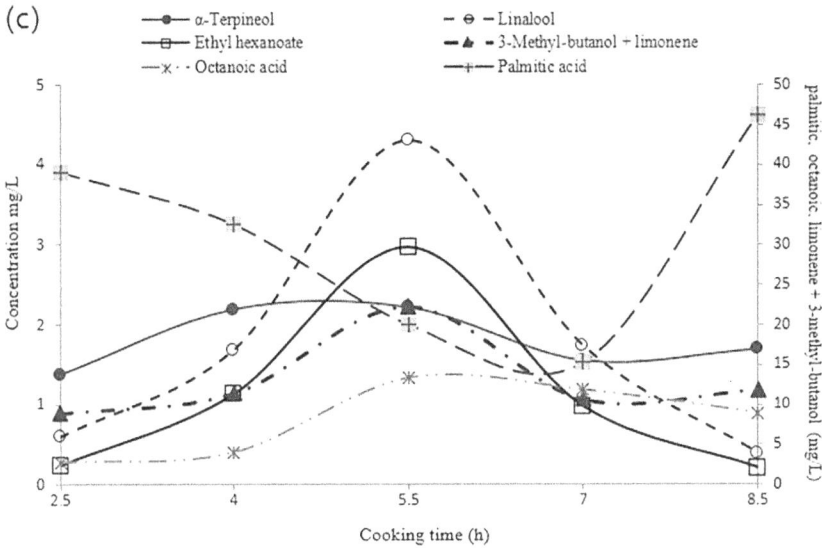

(c)

Legend:
- —●— α-Terpineol
- —□— Ethyl hexanoate
- —×·· Octanoic acid
- — ⊖ —Linalool
- —▲— 3-Methyl-butanol + limonene
- —+— Palmitic acid

Y-axis (left): Concentration mg/L
Y-axis (right): palmitic, octanoic, limonene + 3-methyl-butanol (mg/L)
X-axis: Cooking time (h)

Figure 3.3   (Continued)

3.2.3 Fermentation

In the process of Tequila making, fermentation is considered one of the most important stages because in this step the yeasts, mainly *Saccharomyces cerevisiae*, convert the sugars from the wort (agave juice) into ethanol and the other odor-active compounds characteristic of Tequila (Cedeño 1995). The diversity of volatile and nonvolatile compounds generated, and their concentration depend on the fermentation conditions. The availability of nutrients in agave juice, the carbon:nitrogen ratio, temperature, type of fermentation (spontaneous or induced), yeast strain used (commercial *S. cerevisiae*, wild *S. cerevisiae* isolated, the maker's own wild inocula, or prepared mixed cultures), the origin of sugars in the wort (100% agave juice or 51/49 [%] agave juice/other sugars), presence of agave fiber in wort, and fermentation time are some of the key factors in the generation of the compounds that determine the complexity of the fermentative aromas of Tequila (Cedeño 2003; Pinal *et al.* 1997, 2009; Díaz-Montaño *et al.* 2008; Valle-Rodríguez *et al.* 2012; Morán-Marroquín *et al.* 2011; González-Robles, Estarrón-Espinosa, and Díaz-Montaño 2015; Segura-García *et al.* 2015; Hernández-Cortés *et al.* 2016).

After ethanol, the main product of fermentation, the major volatile components quantified in Tequila worts obtained at the laboratory and industrial levels include higher alcohols, esters, aldehydes, acids, and terpenes (Cedeño 2003).

The higher alcohols 3-methylbutan-1-ol, 2-methylbutan-1-ol, 2-methylpropan-1-ol, propan-1-ol, butan-1-ol, and 2-phenylethan-1-ol are produced during the first hours of fermentation from the amino acids initially present in the wort by the Ehrlich pathway. Higher alcohols can also be synthesized anabolically from the sugars in the medium when there is an amino acid limitation (Figure 3.4a). Research on agave juice fermentation has shown that the yeast strain has an important influence on the formation of higher alcohols, reaching concentrations of 50–318 mg/L (Arellano *et al.* 2008; Díaz-Montaño *et al.* 2008; González-Robles, Estarrón-Espinosa, and Díaz-Montaño 2015; Valle-Rodríguez *et al.* 2012; Morán-Marroquín *et al.* 2011; Segura-García *et al.* 2015).

Esters are generated during fermentation by the condensation between the alcohols produced by yeast and the acyl group donated by acetyl CoA and the condensation of fatty acids with alcohol (Mason and Dufour 2000; Cedeño 2003). Ethyl acetate and ethyl lactate are the predominant esters, and their total concentration at the end of fermentation ranges from 6 to 96 mg/L (Figure 3.4b). Isoamyl acetate, ethyl hexanoate, ethyl octanoate, ethyl decanoate, and phenethyl acetate are also esters generated at this stage and detected without using extractive techniques. The presence of non-*Saccharomyces* yeasts contributes to the generation of these compounds (Segura-García *et al.* 2015; González-Robles, Estarrón-Espinosa, and Díaz-Montaño 2015). Esters contribute significantly to the fruity aroma both in worts during the fermentation stage and in distillates in the Tequila industry.

Among the aldehydes, acetaldehyde is the most abundant at the end of fermentation. This compound is generated as an intermediate product of yeast metabolism during the first hours of fermentation. It has been documented that when the available sugars in the wort decrease, the concentration of this compound also decreases, while its conversion to ethanol continues through the glycolytic pathway (Figure 3.4c); although it is also a precursor of acetoin and ethyl acetate (Cosío Ramirez *et al.* 2002). The concentrations of this compound have been detected in ranges of 2–157 mg/L using different yeast strains and fermentation conditions (Arellano *et al.* 2008, Díaz-Montaño *et al.* 2008; Amaya-Delgado *et al.* 2013; González-Robles, Estarrón-Espinosa, and Díaz-Montaño 2015; Valle-Rodríguez *et al.* 2012; Morán-Marroquín *et al.* 2011).

Another major volatile compound identified in Tequila worts is methanol. Although this compound is generated mainly in the cooking stage, some kinetics have shown an increase of methanol in the course of fermentation (Figure 3.4d). This behavior has been attributed to the action of the pectin-methyl-esterase enzymes present in native or inoculated yeasts that participate in Tequila's fermentation (Tellez Mora 2000).

(a)

(b)

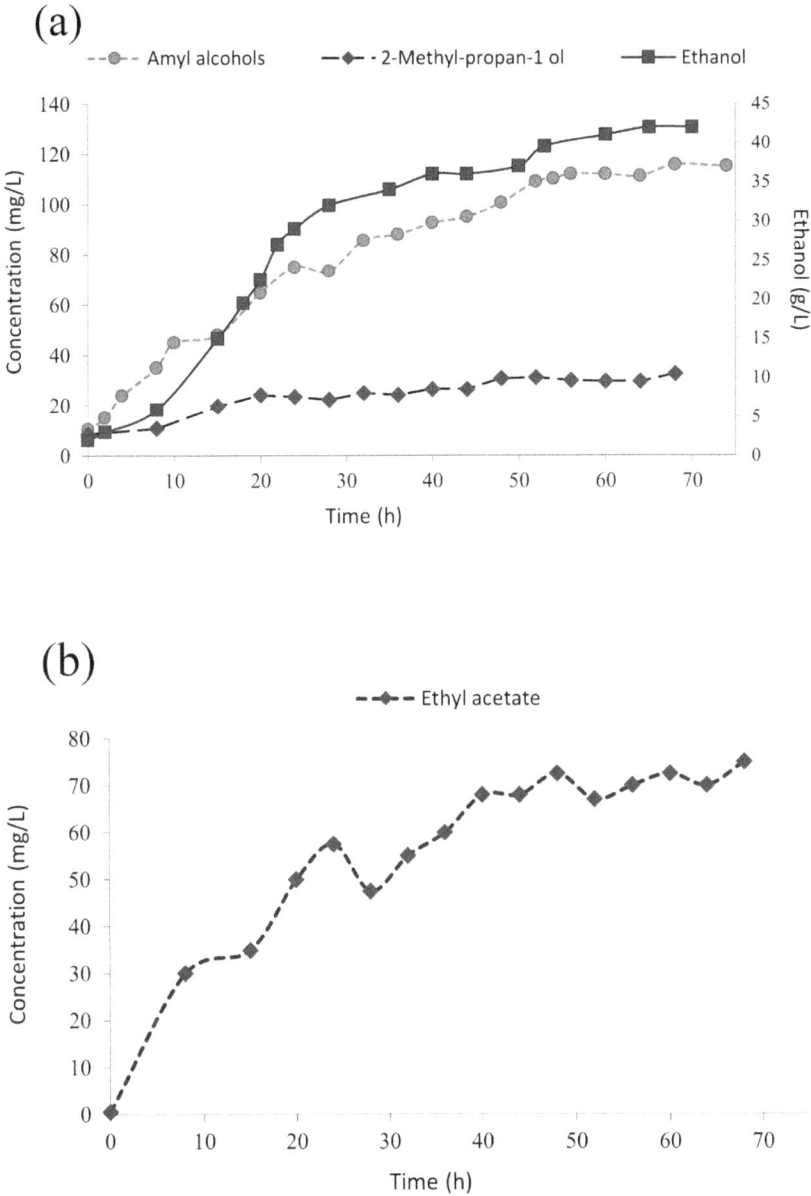

Figure 3.4   Tendencies in the Generation of Volatile Compounds during Agave Juice Fermentation.

(c)

(d)

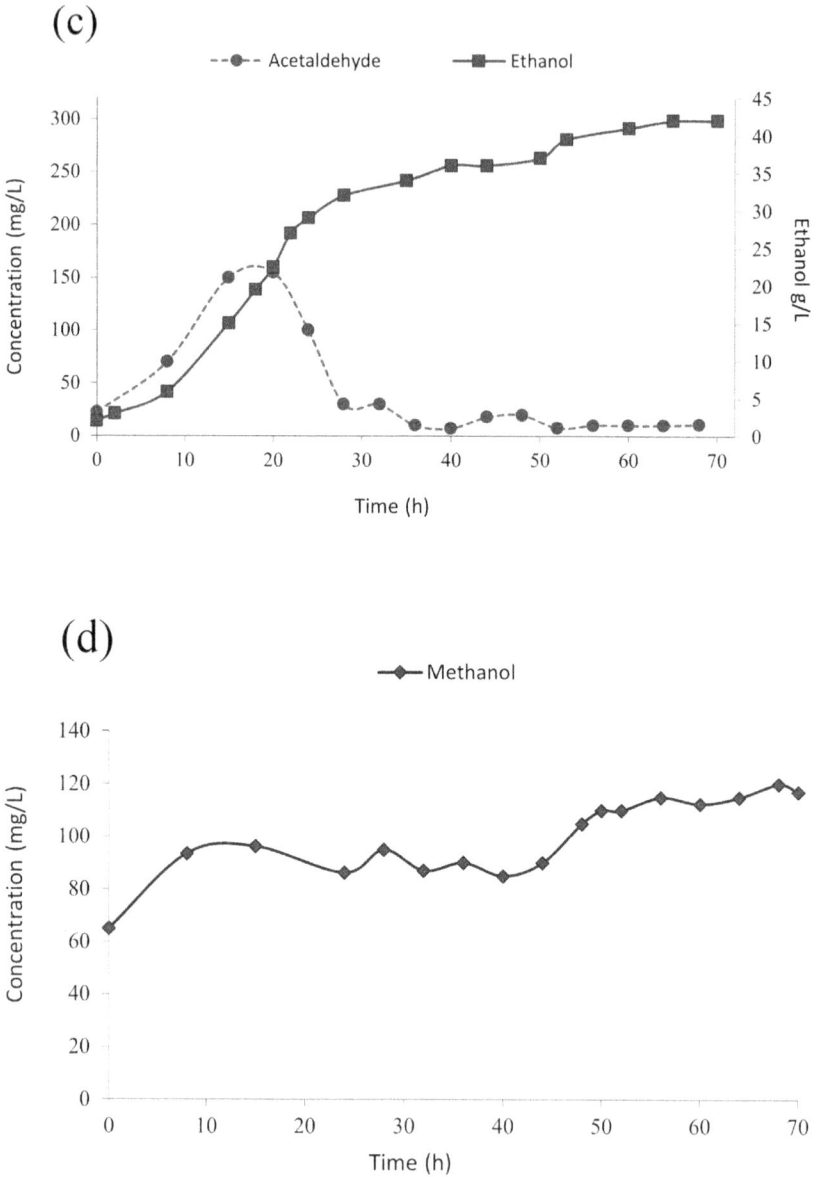

Figure 3.4   (Continued)

On the other hand, volatile organic acids from C2 to C18 have been reported in fermented Tequila worts (Cosío Ramirez *et al.* 2002). Although acetic acid concentration is relatively low in Tequila compared to other distilled beverages made from agave such as mezcal and bacanora, the highest detection of this acid occurs in fermentations carried out with traditional processes. This is attributed to the microbial diversity of the wort, which includes the presence of some bacteria and non-*Saccharomyces* yeasts capable of producing acetic acid through the oxidation of ethanol and acetaldehyde, respectively. Furthermore, it has been documented that long-chain acids are the product of the synthesis of membrane structures during cell growth or the lysis of yeasts at the end of fermentation (Cedeño 2003). Among the organic acids identified in the fermentation of Tequila, acetic, octanoic, decanoic, dodecanoic, hexadecanoic, octadecanoic, oleic, and linoleic acids predominate.

The generation of terpenic compounds has been little studied in Tequila fermentation. The main terpenes identified in Tequila wort come from the raw material as previously described. However, its generation from glycosylated precursors has been suggested due to the hydrolytic activity of the enzymes β-glycosidase, β-cellobiosidase, and β-xylosidase present in non-*Saccharomyces* yeasts and with less activity in *S. cerevisiae* (Arrizon and Gschaedler 2007; Álcazar Valle 2011). Worts characterized in a traditional fermentation showed a similar trend in the major terpenes a-terpineol, linalool, and *cis-/trans*-linalool oxide, which increased their concentration during the first hours of fermentation to decrease later and remain until the end time (Figure 3.5d). This was attributed to the possible presence of non-*Saccharomyces* yeasts in the first hours of fermentation capable of carrying out this conversion, and later, the possible inhibition of their hydrolytic enzymes as ethanol's concentration increased as has been observed in worts of grape. Other unconfirmed pathways for the generation of terpenes in Tequila fermentation could be monoterpenes' bioconversion and the MVA route in *S. cerevisiae*, as in other alcoholic beverages (King and Richard Dickinson 2000).

On the other hand, the furans 5-hydroxymethylfurfural (5-HMF) and furfural, detected in high concentration at the beginning of the fermentation, decrease rapidly during the first hours until reaching levels below the detection limits, as seen in the Figures 3.5e and 3.5f (Cosío Ramirez *et al.* 2002). This tendency was reported by Flores-Cosío *et al.* (2018) in a synthetic medium, explaining this behavior to the detoxification ability of the yeasts *S. cerevisiae* and *K. marxianus*, to convert 5-HMF and furfural into less toxic compounds. Figure 3.5 shows some changes in the generation of volatile compounds during the fermentation of agave juice.

The main volatile components generated during the fermentation of agave juice have been directly evaluated using the HS-GC-FID technique. However, the use of liquid-liquid extraction techniques or SPME followed by GC-MSD has allowed the detection of the enormous volatile wealth in worts from fermentations carried out at both the laboratory and the industrial levels. Other minor compounds identified in this stage by these techniques are indicated in Table 3.1.

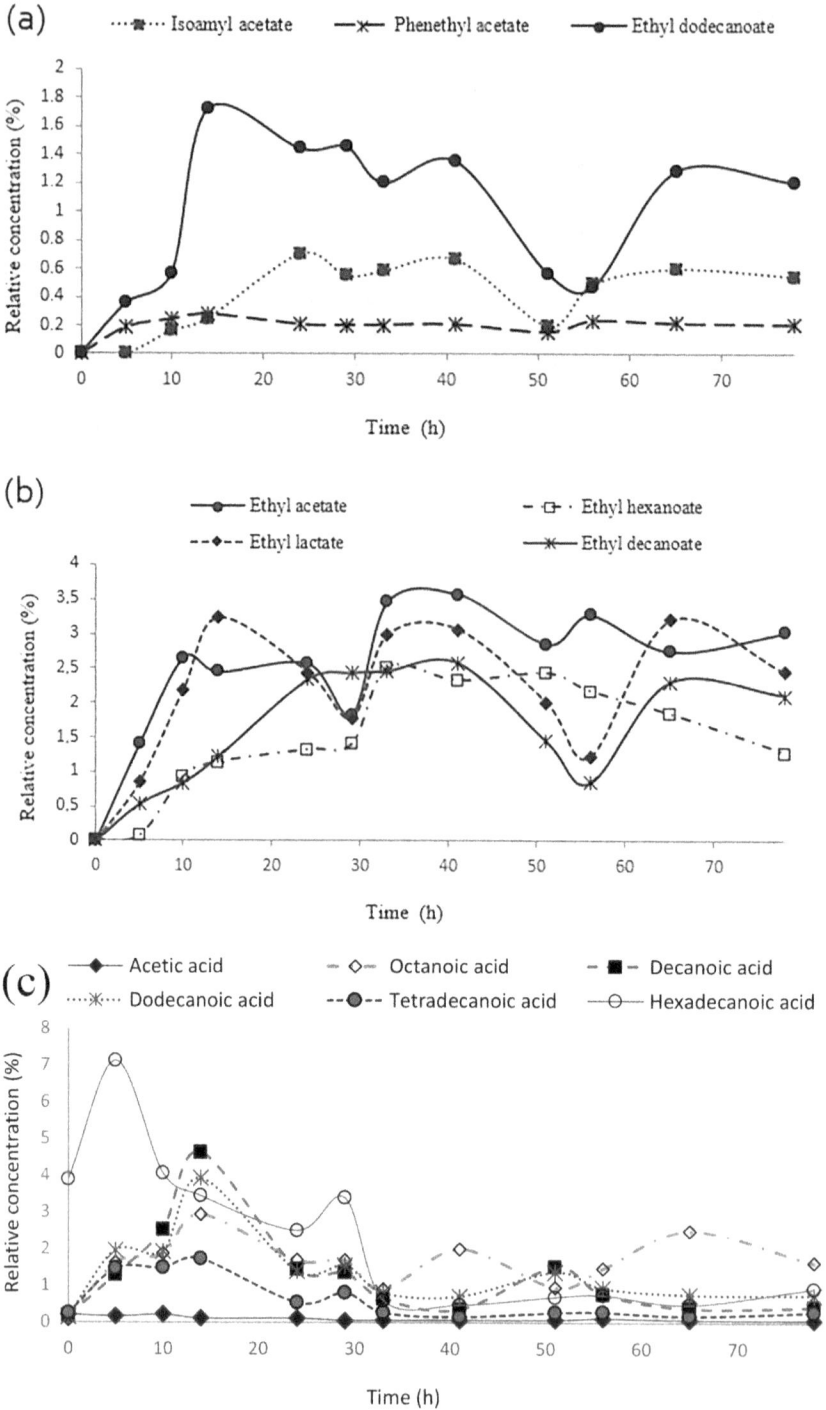

Figure 3.5   Changes in the Generation of Volatile Compounds during Fermentation.

(d)

(e)

(f)

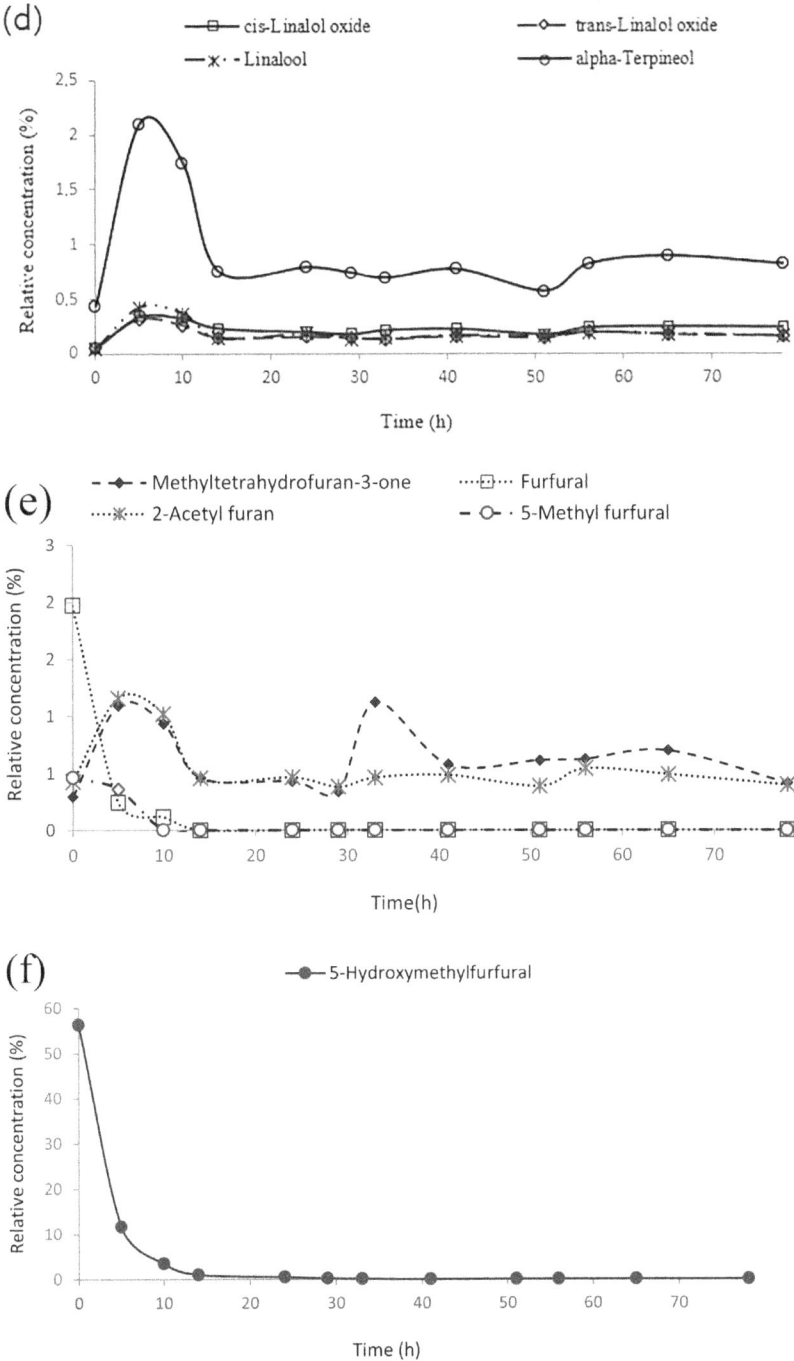

Figure 3.5   (Continued)

3.3 DISTILLATION

Once distillation ends, the resulting wort is rich in ethanol and in a wide variety of volatile compounds in different concentrations. Distillation aims to separate and concentrate the ethanol and preserve the volatile richness generated during the previous steps. Thus an adequate fraction separation program during distillation in still pots is very important in defining the Tequila's sensorial character and accomplishing the physicochemical parameters described in the Official Standard. Based on this, the distillation practice in the Tequila industry is focused on obtaining the highest concentration of ethanol, controlling the maximum allowed limits of its major congeners: higher alcohols (3-methylbutan-1-ol, 2-methylpropan-1-ol, the highest), methanol, esters (ethyl acetate and ethyl lactate), acetaldehyde, and furfural.

A classification of the different types (Type 1→Type 8) of the majority and minority volatile components according to the fraction of the distillate in which they are detected in Tequila has been proposed by Prado Ramírez (2015). This classification considers the boiling temperature and compound's solubility in ethanol and water. A "heads compound" has a higher concentration at the beginning of the distillation (measured in the distillate flow). A "heart and tails compound" presents a higher concentration in the second half of the heart and towards the end of distillation, while a "tails compound" is recovered at the end of distillation. Accordingly, ethanol, acetaldehyde, ethyl acetate, 3-methylbutan-1-ol and 2-methylpropan-1-ol predominate in the heads fraction (Figure 3.6a).

Methanol, acetic acid, ethyl lactate, and furfural are concentrated in the tails fraction (Figure 3.6b). Additionally, in this fraction, medium-, and long-chain organic acids such as octanoic, decanoic, dodecanoic, tetradecanoic, hexadecanoic acids and their ethyl esters are concentrated. Other furan compounds, long-chain alcohols, and phenols also increase in this fraction. The presence of these latter compounds is associated with the prevalence of yeast debris, low sugar concentrations, waxes, solids, or fibers of the agave, as well as minerals and metals (copper [Cu] of some still pot elements). These elements could catalyze other reactions during heating to separate the *ordinario*, while lysis of the yeast remains could also occur.

Finally, the heart, the most important fraction in distillation, contains ethanol and the major congeners in the concentrations allowed by the Official Standard: acetaldehyde, methanol, furfural, esters (ethyl acetate and ethyl lactate), and the higher alcohols (2-butanol, 1-propanol, 2-methylpropan-1-ol, butan-1-ol, 2-methylbutan-1-ol, 3-methylbutan-1-ol, and pentan-1-ol). Moreover, a wide variety of minor compounds, added throughout Tequila's process, confers to the distillate its final aroma and flavor characteristic. The mean

(a)

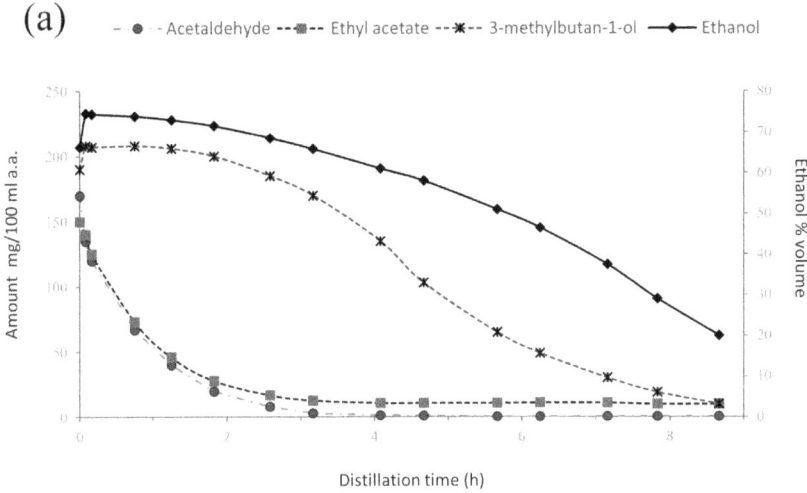

(b)

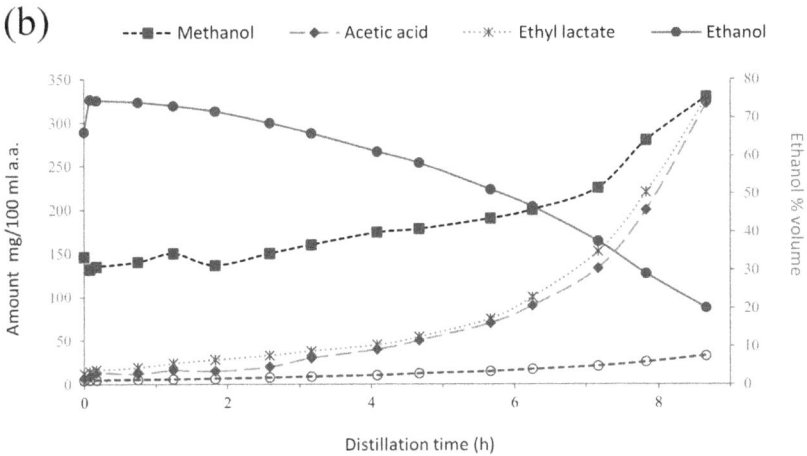

Figure 3.6 Tendency of Volatile Compounds during the Second Distillation in the Tequila Process.

Source: Adapted from Prado Ramírez (2015).

proportion and the detail of the minor volatile fraction in Tequila distillates is shown in Figure 3.7 and Table 3.1, respectively.

The most important factors affecting the volatile profile of distillate have been reported: the design and material of the still pot, the chemical characteristics of the fermented wort or *ordinario*, the volume of the liquid to be distilled,

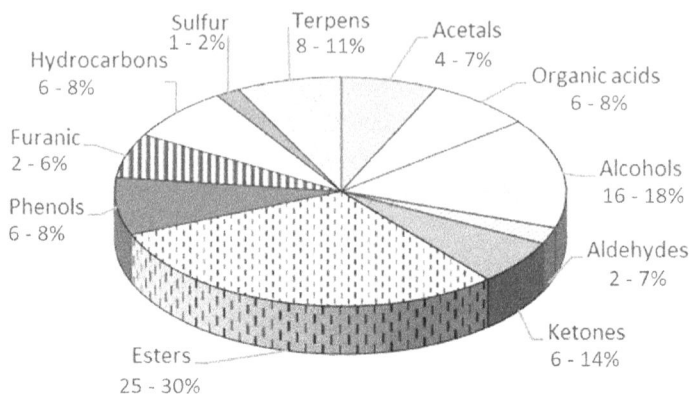

Figure 3.7 Qualitative Proportion of Minor Volatile Compounds in the Heart Fraction.

alcoholic content, the speed of heat supply, as well as the heads- and tails-cutting program during stripping and rectification, among others.

3.4 AGING

Aging or maturation, the final and optional step in Tequila production, is a slow process that makes possible for Tequila to acquire additional sensory characteristics. According to the Official Standard Norm (NOM-006-SCFI-2012 2012), maturation must be carried out in holm oak containers (*Quercus ilex*) or white oak containers (*Quercus alba*). According to the Tequila type, the residence time varies, as shown in Figure 3.1 (NOM-006-SCFI-2012 2012).

During maturation, Tequila acquires sensorial characteristics depending on alcohol content at entry and residence time (López-Ramírez *et al.* 2013; Martín-del-Campo, López-Ramírez, and Estarrón-Espinosa 2019). Additionally, the wood origin (López-Ramírez *et al.* 2013; Martín-del-Campo, López-Ramírez, and Estarrón-Espinosa 2019), cask characteristics (the stave quality and thickness, number of use cycles, regeneration process, toasting degree, capacity, etc.) (Aguilar-Méndez *et al.* 2017; Warren-Vega *et al.* 2021), as well as the cellar conditions (temperature and humidity) have an important impact in the final sensory characteristics of aged Tequilas.

Even though maturation has an important impact on Tequila's aromatic quality, the volatile compounds' profile and their concentrations are not the only factors to evolve. Figure 3.8 shows the evolution of some physicochemical parameters and the volatile compounds regulated in the Official Standard during the maturation in new French oak barrels (López-Ramírez *et al.* 2013). Significant differences among barrel origins have been reported; nevertheless, the tendency is the same for a determined parameter (López-Ramírez *et al.* 2013). One of the most evident changes is the decrease in the light transmittance (Figure 3.8a). At the same time, the dry extract increases (Figure 3.8b). The beverage goes from a transparent color in White Tequila to acquire an amber color in most Aged Tequilas.

Different behaviors are observed depending on the evaluated parameter (López-Ramírez *et al.* 2013). Total acidity and aldehydes content present an important reduction after the first two weeks in barrels, then they increase (Figures 3.8a and 3.8d). Higher alcohols, methanol, esters, and furfural increase during maturation (Figures 3.8a, 3.8c and 3.8d). While higher alcohols rise after two weeks and remain constant (Figure 3.8c), methanol and furfural increase continuously (Figures 3.8c and 3.8d). Recently, Warren-Vega *et al.* (2021) reported that during the tequila maturation process for 12 months using new French oak barrels (New-Cask) and treated with a medium roast (Cask-MT), no significant changes were observed in the concentration of methanol and higher alcohols. However, aldehydes and esters increased during this process.

The increase of higher alcohols has been associated with the ester hydrolysis or beverage evaporation during the residence time (López-Ramírez *et al.* 2013), while the methanol increase could be associated with extraction from the wood since this compound could be produced after the wood toasting by lignin pyrolysis (Fessenden, Fessenden, and Logue 1998). On the other hand, furfural and other furanic compounds are formed during the cooking of *piñas*, but their concentrations increase during maturation mainly due to their extraction from wood. These compounds are formed during barrel toasting by thermal degradation of hemicelluloses, but the extraction rate depends on the alcohol content of the beverage, and the contact time (López-Ramírez *et al.* 2013). Esters generation during the maturation of tequila could be explained by oxidation reactions of polyphenols and flavonoids extracted from the wood due to the oxygen incorporated into the distillate through the barrel's wood (Warren-Vega *et al.* 2021). Its formation by reactions between ethanol and acetic acid in an acidic medium has also been described by what is observed in other alcoholic beverages (López-Ramírez *et al.* 2013).

The aromatic complexity of Aged Tequilas is due to the major and regulated compounds and various volatile compounds belonging to different chemical families (Table 3.1). Of the 485 minor volatile compounds reported in Table 3.1, about 225 compounds have been reported after Tequila has had contact with wood containers. Volatile compounds' evolution during maturation is associated with different physicochemical phenomena. Some compounds are extracted or absorbed from wood, produced due to oxidation or condensation reactions.

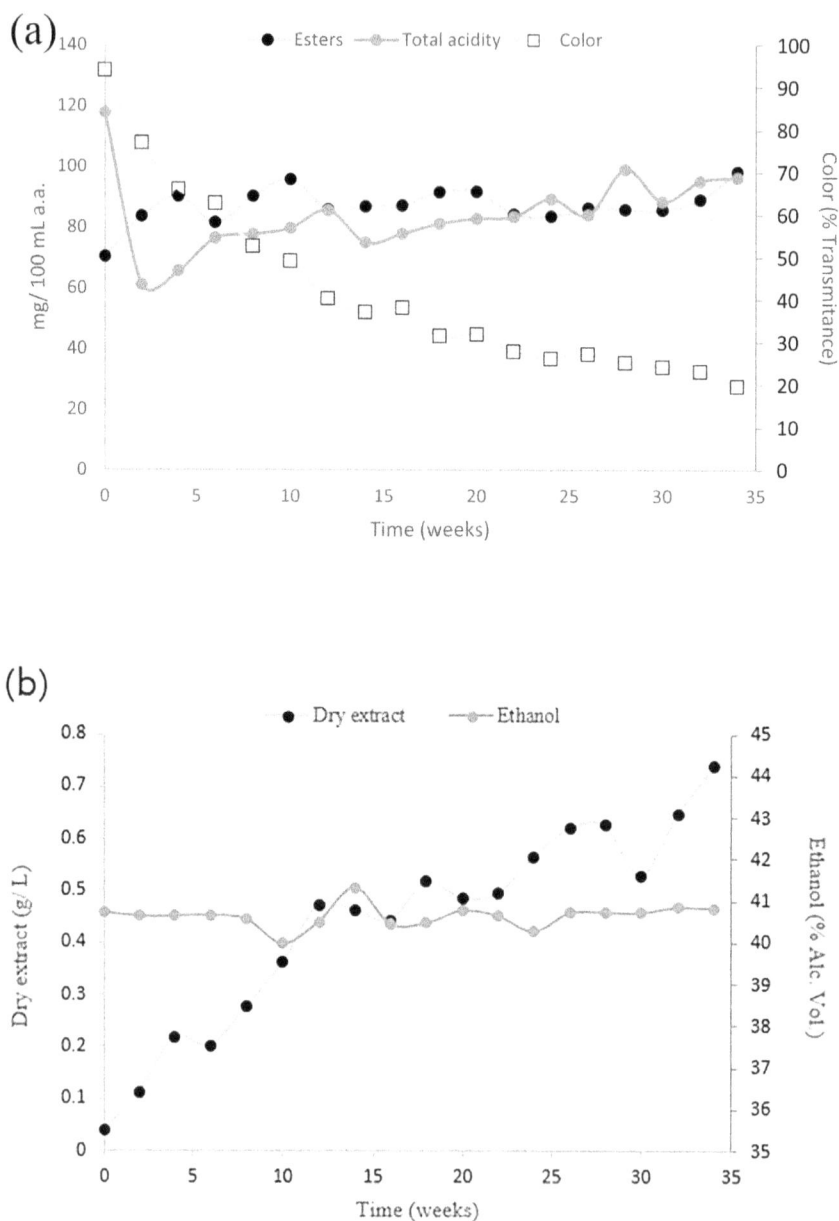

Figure 3.8 Evolution of Major Volatile Compounds and Physicochemical Parameters throughout Maturation.

Source: Adapted from López-Ramírez *et al.* (2013).

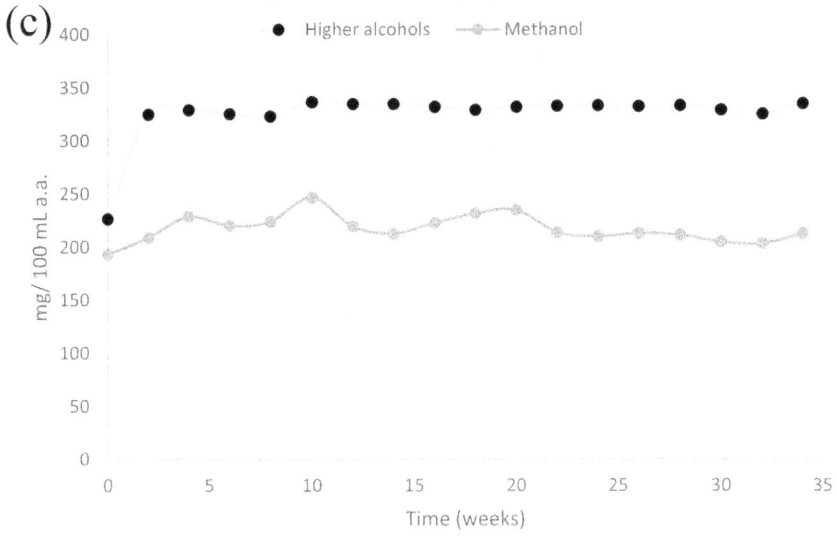

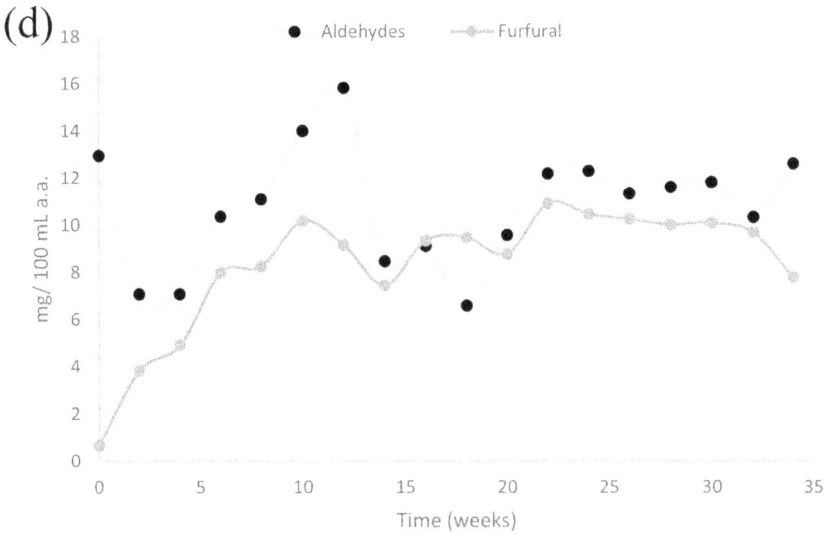

Figure 3.8   (Continued)

Some compounds present in White Tequila disappear during this process, such as 1,1-Diethoxy-2-methyl-butane and 1,1-Diethoxy-3-methyl-butane (Aguilar-Méndez *et al.* 2017). Others appear after two months of maturation, in Aged Tequilas, such as 1,1,1-triethoxypropane, 1,1,1-triethoxyethane, 2-(dieth oxymethyl)furan, 2-methylbutanoic acid, 2,6,6-trimethylcyclohexene-1-carbalde hyde, 2-phenylacetaldehyde, 4-propan-2-ylcyclohexene-1-carbaldehyde, ethyl 2-methyl butanoate, methyl nonanoate, (5-formylfuran-2-yl)methyl acetate, 1-methyl-4-propan-2-ylcyclohex-3-en-1-ol, 1,2,3,4,4a,5,6,7,8,8a-decahydronaph-thalene, 4,4-dimethyl-8-methylene-1-oxaspiro[2.5]octane, 3-furanacetic acid, 4-hexyl-2,5-dihydro-2,5-dioxo, 4,4-dimethyl-8-methylene-1-oxaspiro[2.5]octane (Table 3.1). Others appear only after several months of residence; in Extra-Aged Tequilas, for example, esters such as 3-methyl-butyl pentadecanoate and ethyl tetradecanoate (Aguilar-Méndez *et al.* 2017) or lactones such as *cis-* and *trans-*whisky lactones appear (González-Robles and Cook 2016; Mejia Diaz *et al.* 2019).

Some studies have evaluated how this complex evolution of the volatile compounds profile distinguishes the different Tequila classes. Some compounds have been identified as maturation markers (Mejia Diaz *et al.* 2019; Martín-del-Campo, López-Ramírez, and Estarrón-Espinosa 2019) or as barrel wood origin indicators (Martín-del-Campo, López-Ramírez, and Estarrón-Espinosa 2019) or barrel treatment (Warren-Vega *et al.* 2021). Mejia Diaz *et al.* (2019) reported significant differences between guaiacol, apocynin, and vanillin concentration for White, Aged, or Extra-Aged Tequilas. On the other hand, these authors mentioned that for 4-methylguaiacol, 4-ethylguaiacol, *trans-*whisky lactone, *cis-*whisky lactone, eugenol, and syringaldehyde, no significant differences were observed between Aged or Extra-Aged Tequilas. Martín-del-Campo, López-Ramírez, and Estarrón-Espinosa (2019), described eight compounds showing significant statistical differences between barrel origins and maturation time using new oak barrels: 3-methyl butyl acetate, furan-2-carboxaldehyde, 5-methyl-2-furancarbaldehyde, propanoic acid, 2-methylbutanoic acid, octanoic acid, 1-decanol, and 2,6-dimethoxyphenol. These authors describe compounds that were not present in matured Tequilas depending on the barrel origin (Allier, Limousin, Tronçais, and Centre de la France) used for Tequila maturation. They mention that 1,1-diethoxyisobutane and cyclohex-2-en-1-ol were not identified in Allier, methyl nonanoate in Limousin, 1,2,3,4,4a,5,6,7,8,8a-decahydronaphthalene in Tronçais, and finally 1-phenylpropyl acetate, methyl hexadecanoate, and 2-methoxyphenol in Centre de la France.

3.5 TEQUILA COMPOSITION AND QUALITY

Tequila quality is constructed from the raw material selection and from attention to the operation throughout the whole production process. At the end, this product contains a wide variety of volatile and nonvolatile compounds conferring its unique sensory characteristics (García-Barrón, Villanueva-Rodriguez, and Escalona-Buendía 2021). As part of the quality, the minimum and maximum limits for its major congeners have been established (NOM-006-SCFI-2012 2012). Limits for ethanol content (35–55% alc. vol.), higher alcohols (20–500 mg/100 mL a.a.), methanol (30–300 mg/100 mL a.a.), aldehydes (0–40 mg/100 mL a.a.), and furfural (0–4 mg/100 mL a.a.) are the same for all the Tequila classes. The allowed limits concerning dry extract for White Tequila (0–0.3 g/L) are lower than for Gold, Aged, and Extra-Aged Tequilas (0–5 g/L). On the other hand, the same limit for esters content (2–200 mg/100 mL a.a.) is allowed for White and Gold Tequila, while, for Aged and Extra-Aged Tequilas 2–250 mg/100 mL a.a. are allowed.

Additionally, Tequila's global aroma is enriched with a wide variety of minor volatile compounds. Recent investigations have reported the identification of more than 590 volatile compounds (Rodríguez-Olvera *et al.* 2019). The amount, type, and proportion of those minor volatile compounds, confer the sensory characteristics whereby the final product satisfies consumer preferences.

3.6 CONCLUSION

Tequila is a traditional distilled beverage with special characteristics. Its complex sensory attributes, which are acquired through the generation of aromatic compounds during the production process, are appreciated by consumers in products (such as White Tequila) that are just distilled. This important characteristic differentiates Tequila from other more neutral distillates just obtained from the alembic. White Tequila presents a complex aroma profile derived from compounds that originate not only during fermentation but in the previous steps, such as the raw material selection and the cooking of the *piñas*.

Even though there is published information about Tequila's volatile composition and the impact of the process stages on the quality of the final product in controlled processes at the laboratory scale, it is necessary to increase the cooperation between academia and the industrial sector to perform deeper research on an industrial scale on the phenomena that give rise to the sensory complexity of this beverage in traditional processes and those that involve the use of diffusers or a combination of both.

REFERENCES

Aguilar-Méndez, O., J. A. López-Álvarez, A. L. Díaz-Pérez *et al.* 2017. Volatile compound profile conferred to tequila beverage by maturation in recycled and regenerated white oak barrels from *Quercus alba*. *European Food Research and Technology* 43 (12):2073–2082. doi:10.1007/s00217-017-2901-7.

Álcazar Valle, E. M. 2011. *Capacidades fermentativas y generación de volátiles de cepas de levaduras aislaldas en diferentes estados productores de mezcal*. Master, CIATEJ. https://ciatej.repositorioinstitucional.mx/jspui/bitstream/1023/91/1/Elba%20 Montserrat%20Alcazar%20Valle.pdf.

Amaya-Delgado, L., E. J. Herrera-López, J. Arrizon, M. Arellano-Plaza, and A. Gschaedler. 2013. Performance evaluation of *Pichia kluyveri*, *Kluyveromyces marxianus* and *Saccharomyces cerevisiae* in industrial tequila fermentation. *World Journal of Microbiology and Biotechnology* 29 (5):875–881. doi:10.1007/s11274-012-1242-8.

Arellano, M., C. Pelayo, J. Ramírez, and I. Rodriguez. 2008. Characterization of kinetic parameters and the formation of volatile compounds during the tequila fermentation by wild yeasts isolated from agave juice. *Journal of Industrial Microbiology & Biotechnology* 35 (8):835–841. doi:10.1007/s10295-008-0355-4.

Arrizon, J., and A. Gschaedler. 2007. Effects of the addition of different nitrogen sources in the tequila fermentation process at high sugar concentration. *Journal of Applied Microbiology* 102 (4):1123–1131. doi:10.1111/j.1365-2672.2006.03142.x.

Benn, S. M., and T. L. Peppard. 1996. Characterization of tequila flavor by instrumental and sensory analysis. *Journal of Agricultural and Food Chemistry* 44 (2):557–566. doi: 10.1021/jf9504172.

Cedeño, M. 1995. Tequila production. *Critical Reviews in Biotechnology* 15 (1):1–11. doi:10.3109/07388559509150529.

Cedeño, M. 2003. Tequila production from agave: Historical influences and contemporary processes. In *The Alcohol Textbook: A Reference for the Beverage, Fuel and Industrial Alcohol Industries*, edited by K. A. Jaques, T. P. Lyons, and D. R. Kelsall, 223–245. Nottingham: Nottingham University Press.

CIATEJ. 1999. *Identificación de la huella distintiva en clases y tipos de Tequila de 44 fábricas de la región para la determinación de adulteraciones*. Reporte técnico final. [Identification of the distinctive imprint in classes and types of Tequila from 44 factories in the region for the determination of adulterations. Final technical report]. SIMORELOS, Clave 96-05-004.

Cosío Ramirez, R., M. Estarrón Espinosa, S. T. Martín del Campo Barba *et al.* 2002. *Estudio sobre la Evolución de los compuestos organolépticos durante el proceso de elaboración del tequila y su relación con la calidad del producto. Informe técnico final* [*Study on the Evolution of Organoleptic Compounds During the Process of Making Tequila and Its Relationship with the Quality of the Product. Final Technical Report*]. Guadalajara: SIMORELOS.

Díaz-Montaño, D. M., M. L. Délia, M. Estarrón-Espinosa, and P. Strehaiano. 2008. Fermentative capability and aroma compound production isolated from *Agave tequilana* Weber juice. *Enzyme and Microbial Technology* 42 (7):608–616. doi:10.1016/j.enzmictec.2007.12.007.

Estarrón, M. 1997. *Identificación de los principales compuestos volátiles que caracterizan al tequila 100% de agave* [IIdentification of the Main Volatile Compounds that Characterize 100% Agave Tequila]. Maestría en Ciencias en Procesos Biotecnológicos Maestría en Ciencias, Ingeniería Química, Universidad de Guadalajara.

Estarrón Espinosa, M., S. T. Martín del Campo Barba, and L. Pinal Suazo. 2001. *Cuantificación de los compuestos organolépticos en fracciones de cocimiento y fermentación obtenidos a nivel laboratorio. Informe técnico* [*Quantification of Organoleptic Compounds in Cooking and Fermentation Fractions Obtained at the Laboratory Level. Technical Report*]. Guadalajara: CIATEJ.

Fessenden, R. J., J. S. Fessenden, and M. W. Logue. 1998. *Organic Chemistry*, 6th ed. Pacific Grove: Brooks, Cole Pub Co.

Flores-Cosío, G., M. Arellano-Plaza, A. Gschaedler, and L. Amaya-Delgado. 2018. Physiological response to furan derivatives stress by Kluyveromyces marxianus SLP1 in ethanol production. *Revista Mexicana de Ingeniería Química* 17 (1):189–202.

García-Barrón, S. E., S. Villanueva-Rodriguez, and H. Escalona-Buendía. 2021. Mezcal and tequila: Volatile composition and sensory characteristics of two traditional beverages. *Fermented and Distilled Alcoholic Beverages: A Technological, Chemical and Sensory Overview. Distilled Beverages*: 125–155.

González-Robles, I. W., and D. J. Cook. 2016. The impact of maturation on concentrations of key odour active compounds which determine the aroma of tequila. *Journal of the Institute of Brewing* 122 (3):369–380. doi:10.1002/jib.333.

González-Robles, I. W., M. Estarrón-Espinosa, and D. M. Díaz-Montaño. 2015. Fermentative capabilities and volatile compounds produced by Kloeckera/ Hanseniaspora and Saccharomyces yeast strains in pure and mixed cultures during Agave tequilana juice fermentation. *Antonie van Leeuwenhoek* 108 (3):525–536. doi:10.1007/s10482-015-0506-3.

Hernández-Cortés, G., J. O. Valle-Rodríguez, E. J. Herrera-López *et al.* 2016. Improvement on the productivity of continuous tequila fermentation by Saccharomyces cerevisiae of Agave tequilana juice with supplementation of yeast extract and aeration. *AMB Express* 6 (1):47. doi:10.1186/s13568-016-0218-8.

King, A., and J. Richard Dickinson. 2000. Biotransformation of monoterpene alcohols by Saccharomyces cerevisiae, Torulaspora delbrueckii and Kluyveromyces lactis. *Yeast* 16 (6):499–506. doi:10.1002/(SICI)1097-0061(200004)16:6 < 499: :AID-YEA548 > 3.0.CO;2-E.

Lopez, M. G. 1999. Tequila aroma. In *Flavor Chemistry of Ethnic Foods*, edited by F. Shahidi, and C. T. Ho, 211–217. New York: Kluwer Academic/Plenum Publishers.

Lopez, M. G., and J. P. Dufour. 1999. Charm analyses of Blanco, Reposado, and Anejo tequilas. *Abstracts of Papers of the American Chemical Society* 218:79-AGFD.

López-Ramírez, J. E., S. T. Martín-del-Campo, H. Escalona-Buendía, J. A. García-Fajardo, and M. Estarrón-Espinosa. 2013. Physicochemical quality of tequila during barrel maturation. A preliminary study. *CyTA—Journal of Food* 11 (3):223–233. doi:10.108 0/19476337.2012.727033.

López-Romero, J. C., J. F. Ayala-Zavala, G. A. González-Aguilar, E. A. Peña-Ramos, and H. González-Ríos. 2018. Biological activities of Agave by-products and their possible applications in food and pharmaceuticals. *Journal of the Science of Food and Agriculture* 98 (7):2461–2474.

López-Salazar, H., B. H. Camacho-Díaz, S. V. Ávila-Reyes *et al.* 2019. Identification and quantification of β-Sitosterol β-d-Glucoside of an ethanolic extract obtained by microwave-assisted extraction from agave angustifolia haw. *Molecules* 24 (21):3926.

Magana, Armando Alcazar, Kazimierz Wrobel, Julio Cesar Torres Elguera, Alma Rosa Corrales Escobosa, and Katarzyna Wrobel. 2015. Determination of small phenolic compounds in tequila by liquid chromatography with ion trap mass spectrometry detection. *Food Analytical Methods* 8 (4):864–872. doi: 10.1007/s12161-014-9967-7.

Mancilla-Margalli, N. A., and M. G. López. 2002. Generation of Maillard compounds from inulin during the thermal processing of *Agave tequilana* Weber var. azul. *Journal of Agricultural and Food Chemistry* 50 (4):806–812. doi:10.1021/jf0110295.

Martín-del-Campo, S. T., J. E. López-Ramírez, and M. Estarrón-Espinosa. 2019a. Dataset of volatile compounds identified, quantified and GDA generated of the maturation process of silver tequila in new French oak barrels. *Data in Brief* 27:104707. doi: 10.1016/j.dib.2019.104707.

Martín-del-Campo, S. T., J. E. López-Ramírez, and M. Estarrón-Espinosa. 2019b. Evolution of volatile compounds during the maturation process of silver tequila in new French oak barrels. *LWT* 115:108386. doi: 10.1016/j.lwt.2019.108386.

Mason, A. B., and J. P. Dufour. 2000. Alcohol acetyltransferases and the significance of ester synthesis in yeast. 16 (14):1287–1298. doi:10.1002/1097-0061(200010) 16:14 < 1287::AID-YEA613 > 3.0.CO;2-I.

Mejia Diaz, L. F., K. Wrobel, A. R. Corrales Escobosa, D. A. Aguilera Ojeda, and K. Wrobel. 2019. Identification of potential indicators of time-dependent tequila maturation and their determination by selected ion monitoring gas chromatography—mass spectrometry, using salting-out liquid—liquid extraction. *European Food Research and Technology* 245 (7):1421–1430. doi:10.1007/s00217-019-03271-7.

Morán-Marroquín, G. A., J. Córdova, J. O. Valle-Rodríguez, M. Estarrón-Espinosa, and D. M. Díaz-Montaño. 2011. Effect of dilution rate and nutrients addition on the fermentative capability and synthesis of aromatic compounds of two indigenous strains of *Saccharomyces cerevisiae* in continuous cultures fed with *Agave tequilana* juice. *International Journal of Food Microbiology* 151 (1):87–92. doi:10.1016/j. ijfoodmicro.2011.08.008.

NOM-006-SCFI-2012.2012. *Bebidas alcohólicas-Tequila-Especificaciones*. México: Diario Oficial de la Federación.

Peña-Alvarez, A., S. Capella, R. Juarez, and C. Labastida. 2006. Determination of terpenes in tequila by solid phase microextraction-gas chromatography-mass spectrometry. *Journal of Chromatography A* 1134 (1–2):291–297.

Pérez Martínez, F. J., E. Rodríguez González, M. Arellano Plaza, R. M. Camacho Ruiz, and R. Prado Ramírez. 2015. Extracción del jugo de agave. In *Ciencia y Tecnología del Tequila: Avances y Perspectivas*, edited by A. Gschaedler Mathis, Benjamín Rodríguez Garay, R. Prado Ramírez, and José Luis Flores Montaño, 55–95. Guadalajara: CIATEJ.

Pinal, L. 2001. *Influencia del tiempo de cocimiento sobre la generación de compuestos organolépticos en las etapas de cocimiento y fermentación de la elaboración del tequila.* Master. Guadalajara: Ingeniería Química, Universidad de Guadalajara.

Pinal, L., M. Cedeno, H. Gutiérrez, and J. Alvarez Jacobs. 1997. Fermentation parameters influencing higher alcohol production in the tequila process. *Biotechnology Letters* 19 (1):45–47.

Pinal, L., E. Cornejo, M. Arellano *et al.* 2009. Effect of *Agave tequilana* age, cultivation field location and yeast strain on tequila fermentation process. *Journal of Industrial Microbiology and Biotechnology* 36 (5):655–661. doi:10.1007/s10295-009-0534-y.

Prado Ramírez, R. 2015. Destilación. In *Ciencia y Tecnología del Tequila: Avances y Perspectivas*, edited by A. Gschaedler Mathis, Benjamín Rodríguez Garay, R. Prado Ramírez and José Luis Flores Montaño, 181–230. Guadalajara: CIATEJ.

Prado-Jaramillo, N., M. Estarrón-Espinosa, H. Escalona-Buendía, R. Cosío-Ramírez, and S. T. Martín-del-Campo. 2015. Volatile compounds generation during different stages of the Tequila production process. A preliminary study. *LWT—Food Science and Technology* 61:471–483. doi:10.1016/j.lwt.2014.11.042.

Rodríguez, David Muñoz, Katarzyna Wrobel, and Kazimierz Wrobel. 2005. Determination of aldehydes in tequila by high-performance liquid chromatography with 2, 4-dinitrophenylhydrazine derivatization. *European Food Research and Technology* 221 (6):798–802. doi: 10.1007/s00217-005-0038-6.

Rodríguez Garay, B., A. Gutiérrez Mora, J. Arrizon *et al.* 2015. La Materia Prima: *Agave tequilana* Weber Var. Azul. In *Ciencia y Tecnología del Tequila: Avances y Perspectivas*, edited by A. Gschaedler Mathis, Benjamín Rodríguez Garay, R. Prado Ramírez, and José Luis Flores Montaño, 17–53. Guadalajara: CIATEJ.

Rodríguez-Olvera, M., L. Rodríguez-Rodríguez, M. Qian, Y. Qian, and P. Vazquez-Landaverde. 2019. Implementation of stir bar sorptive extraction (SBSE) for the analysis of volatile compounds in Tequila. In *Sex, Smoke, and Spirits: The Role of Chemistry*, edited by Brian Guthrie, Jonathan D. Beauchamp, Andrea Buettner, Stephen Toth, and Michael C. Qian, 311–324. Washington, DC: ACS Publications.

Segura-García, L. E., P. Taillandier, C. Brandam, and A. Gschaedler. 2015. Fermentative capacity of Saccharomyces and non-Saccharomyces in agave juice and semi-synthetic medium. *LWT—Food Science and Technology* 60 (1):284–291. doi:10.1016/j.lwt.2014.08.005.

Sidana, J., B. Singh, and O. P. Sharma. 2016. Saponins of agave: Chemistry and bioactivity. *Phytochemistry* 130:22–46.

Solís-García, A., P. Rivas-García, C. Escamilla-Alvarado, R. Rico-Martínez, M. G. Bravo-Sánchez, and J. E. Botello-Álvarez. 2017. Methanol production kinetics during agave cooking for mezcal industry. *Revista Mexicana de Ingeniería Química* 16 (3):827–834.

Tellez Mora, P. 2000. *Análisis de las variables que influyen en la síntesis de metanol, en la producción de tequila durante las etapas de cocimiento y fermentación*. Master. Guadalajara: Ingeniería Química, Universidad de Guadalajara.

Valle-Rodríguez, J. O., G. Hernández-Cortés, J. Córdova, M. Estarrón-Espinosa, and D. Díaz-Montaño. 2012. Fermentation of agave tequilana juice by *Kloeckera africana*: Influence of amino-acid supplementations. *Antonie Van Leeuwenhoek International Journal of General and Molecular Microbiology* 101 (2):195–204. doi:10.1007/s10482-011-9622-x.

Vallejo-Cordoba, B., A. F. Gonzalez-Cordova, and M. del Carmen Estrada-Montoya. 2004. Tequila volatile characterization and ethyl ester determination by solid phase microextraction gas chromatography/mass spectrometry analysis. *Journal of Agricultural and Food Chemistry* 52 (18):5567–5571. doi: 10.1021/jf0499119.

Waleckx, E., A. Gschaedler, B. Colonna-Ceccaldi, and P. Monsan. 2008. Hydrolysis of fructans from *Agave tequilana* Weber var. azul during the cooking step in a traditional tequila elaboration process. *Food Chemistry* 108 (1):40–48. doi:10.1016/j.foodchem.2007.10.028.

Warren-Vega, W. M., R. Fonseca-Aguiñaga, L. V. González-Gutiérrez, F. Carrasco-Marín, A. I. Zárate-Guzmán, and L. A. Romero-Cano. 2021. Chemical characterization of tequila maturation process and their connection with the physicochemical properties of the cask. *Journal of Food Composition and Analysis* 98:103804. doi:10.1016/j.jfca.2021.103804.

Chapter 4

Volatile Compounds Formation in Whisky

Barry Harrison

CONTENTS

DOI: 10.1201/9781003129462-5

4.1 INTRODUCTION

Whisky is a type of alcoholic beverage produced by the distillation of fermented grains followed by aging in wooden casks. The origins of this beverage are not clear; the earliest documented records are from Scotland and Ireland dating back to the 15th century, but it is likely that unrecorded whisky was being produced prior to that (Master of Malt, n.d.). Scotland and Ireland remain major producers, but whisky is now produced by many other countries around the world, with other major producers including the United States, Japan, and Canada. As the largest producer and as the area of the author's experience, the focus of this chapter will be whisky produced in Scotland—Scotch Whisky.

Within the Scotch Whisky category, there are two types—Grain Whisky and Malt Whisky—which can be differentiated by production process. The majority of Scotch Whisky is sold as blended whisky which is a blend of a number of different grain and malt whiskies. Grain whisky tends to make up the largest proportion of total volume in a blend but the malt whiskies tend to contribute the largest proportion of flavor compounds. Indeed, single malt whisky—malt whisky from a single distillery—is a popular beverage in itself. Given its contribution to flavor, focus will be made here on the flavor compounds found in malt whisky and how they are formed in the production process. Although the emphasis is on Scotch Malt Whisky, many of the compounds and processes will apply to other whiskies.

4.2 SCOTCH MALT WHISKY PRODUCTION PROCESS OVERVIEW

Strict legal provisions govern how Scotch Whisky is made, marketed, and exported. Scotch Whisky has been defined by statute in the United Kingdom since 1933 and the current definition is set out in the Scotch Whisky Regulations 2009 ("the UK Law") (The Scotch Whisky Association, n.d.; UK Parliament 2009). In summary, Scotch Whisky must be:

- Made in Scotland from only cereals, water, and yeast (no flavoring or sweetening is permitted);
- Distilled below 94.8% abv (alcohol by volume) so that it retains the flavor and aroma derived from its raw materials;
- Matured for a minimum of three years in oak casks;
- Bottled at a minimum strength of 40% abv.

The three main raw materials used for malt whisky production are barley, water, and yeast. The production process of malt whisky consists of malting, milling, mashing, fermentation, distillation, and maturation. In-depth technical reviews of the process can be found in the literature; a brief overview is provided here (Russell and Stewart 2014; Piggott, Sharp, and Duncan 1989).

Malting is carried out to modify the physical structure of the barley grain and to ensure that a series of enzymes are in place to allow the subsequent extraction of fermentable sugars. The malting process involves steeping the barley in water to initiate germination then stopping the process by the application of heat in a kiln when an appropriate level of modification is reached. In the distillery, the malt is milled to produce a grist which is mixed with hot water in a mashing vessel. In this vessel, known as a mash tun, the starch in the malt is predominantly converted to maltose as well as other fermentable sugars. The sugary liquid extracted from the malt, known as wort, is filtered from the grist in the mash tun and pumped to the fermentation vessel, or washback.

As the washback starts to fill with wort, yeast is added, and the fermentation process begins. This process is allowed to continue for at least 48 hours and sometimes can be left for 100 hours or more; the product is a simple beer known as wash. The fermented wash is then most commonly double distilled in copper pot stills, though there are some instances of triple distillation. In a typical double distillation, the first distillation serves to concentrate the alcohol from the wash in a distillate known as low wines. The low wines is then redistilled in a second still to further concentrate the alcohol in a fraction of distillate known as new make spirit, which also contains desirable levels of flavor compounds. The new make spirit is the second of three distillate fractions; the first (foreshots) and third (feints) are recycled and added to the next batch of low wines.

New make spirit is then matured in an oak cask to produce whisky. The major sources of casks used by the Scotch Whisky industry are those previously used in the bourbon and sherry industries. Scotch Whisky producers can then reuse the casks several times until they no longer provide the required levels of extractives. At this point the cask will either be taken out of use, or it may be regenerated at the cooperage by removing some of the exhausted inner surface layer and reapplying heat treatment to the newly exposed surface. The maturation period is required to be at least three years but is typically ten years or more for a Scotch Malt Whisky.

4.3 SCOTCH MALT WHISKY FLAVOR

Flavor is, of course, fundamental to a whisky brand identity and determines its success in the marketplace. Producers therefore want to be able to control flavor in the production process. This control is needed to maintain flavor consistency for established brands, but also to provide the knowledge to enable

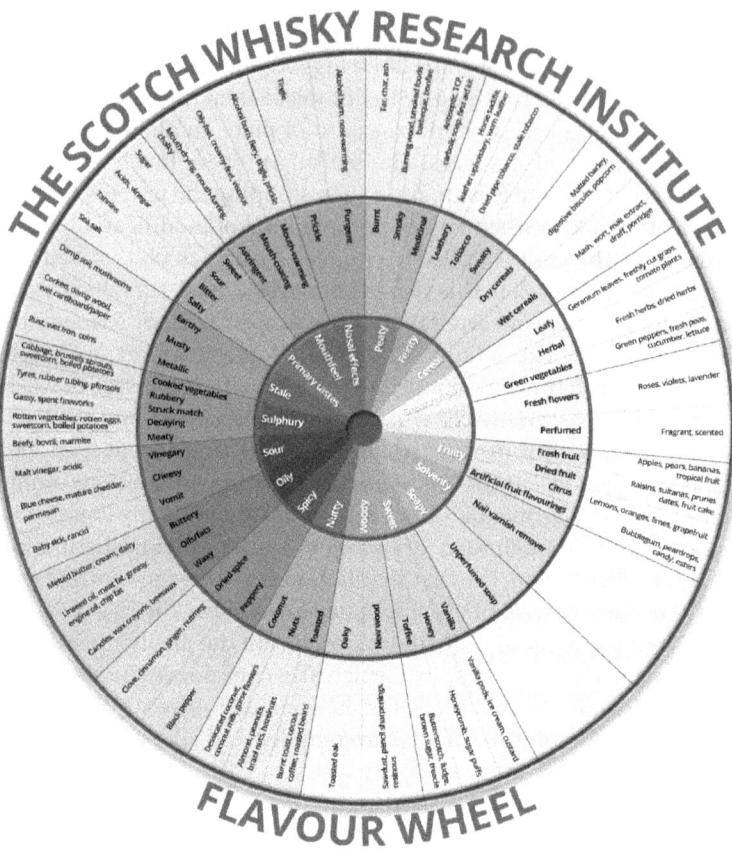

Figure 4.1   The Scotch Whisky Research Institute Flavour Wheel.

innovation for new brands. The key to flavor control is flavor measurement, and the primary way to measure flavor is through sensory assessment.

Whisky flavor is complex and the result of the balance of a large number of flavor attributes. Various sensory approaches can be used to assess flavor, but an invaluable tool to ensure that the communication of complex flavor is reliable is to use a predetermined and agreed-upon vocabulary. This need was addressed for Scotch Whisky in the 1970s with the development of a whisky flavor wheel by industry experts (Shortreed *et al.* 1979). This has evolved to become The Scotch Whisky Research Institute Flavour Wheel (Figure 4.1), which is widely used in the industry. This tool can help categorize more specific flavor descriptors detected by individuals on the outer circle into the more general descriptions on the center circle of the wheel. These general descriptions can then be used to simplify the processes of comparing and quantifying aromas across products.

Once the flavor of a whisky has been defined, the next control challenge is to understand how that flavor was created in the process. In that regard, work has been ongoing for many decades to define the compounds responsible for the various aspects of whisky flavor. Records going back to around the turn of the 20th century document the chemists desire to understand the chemistry of whisky and its flavor (Schidrowitz and Kaye 1905; Schidrowitz 1902).

The most reliable way to understand how well we can define flavor in terms of composition is to recreate the composition of a whisky using our knowledge of the flavor compounds and to determine how well the simulant matches the original whisky. In 1972, Salo *et al.* carried out such an experiment recombining the flavor compounds they could measure in a blended Scotch Whisky (Salo, Nykanen, and Suomalainen 1972). They combined groups of esters, alcohols and organic acids—known to be quantitatively the major groups in whisky—along with a group of carbonyls (Watson 1983b). They found that the simulant aroma did not exactly match the original whisky indicating some compounds were still to be identified and added. However, a key point made by Salo *et al.* was that, of the compound groups included, the carbonyls played the most significant role in the flavor of the whisky despite being present at lower levels. This was a good example of how the impact on aroma is a function of the aroma threshold of a compound as well as its concentration in the whisky. Another useful point raised in that paper was that it is not simply an additive effect of compounds above their aroma threshold that contribute to the overall aroma. They observed, for example, that the odor activity of the combined acid fraction was lower than some of the individual acids, suggesting a suppressive effect in the combined fraction.

The relationship between flavor and composition is complex, with additive, suppressive and synergistic effects of different compounds all playing a role in the overall flavor. We must therefore continue to build our knowledge of the compounds contributing to flavor until we get to a point where we can produce a model spirit fully representative of the real product. Helping us in that challenge are advancements in analytical instruments; we have moved on significantly since the turn of the 20th century (Schidrowitz and Kaye 1905). Miniaturized preconcentration techniques such as solid-phase microextraction (SPME) are now common in analytical laboratories and allow the automated extraction of compounds from small sample volumes. Combining such techniques with sensitive detectors such as mass spectrometers has made the analysis of many trace compounds relatively straightforward. However, some flavor compounds are present at such low levels that they are difficult to detect even using these sensitive techniques. Further developments to improve sensitivity, utilizing novel pre-concentration techniques and ever more powerful detectors, will improve the ability to detect the more elusive trace compounds. In addition, advances in artificial intelligence will likely help account for the combined and synergistic effects of flavor compounds when relating composition to flavor. As the understanding of the compounds contributing to flavor continues to increase, a greater appreciation of how different parts of the production process interact to influence compound levels will provide greater control over the final product flavor profile.

4.4 FORMATION OF FLAVOR COMPOUNDS

4.4.1 Cereal

In malt whisky production the only cereal used is barley. The key aroma compounds of barley have been reported as 3-methylbutanal, 2-methylbutanal, hexanal, 2-hexenal, 2-heptenal, (E)-2-nonenal and decanal (Cramer *et al.* 2005). These compounds can be produced biochemically or chemically, with 3-methyl butanal and 2-methylbutanal being derived from amino acids and hexanal, 2-hexenal, 2-heptenal, (E)-2-nonenal and decanal from fatty acids (Cramer *et al.* 2005; Smit, Engels and Smit 2009). 3-methylbutanal and 2-methylbutanal are considered to have malty aromas (Dong *et al.* 2013; Beal and Mottram 1994). The lipid-derived aldehydes have varied aromas; for example, hexanal and (E)-2-hexenal have grassy type aromas, and (E)-2-nonenal has a fatty or cardboard aroma and plays a well documented role in beer staling (Dong *et al.* 2013; Vanderhaegen *et al.* 2006).

There is little literature on the exact nature of the pathways leading to these compounds formation in barley, but Cramer *et al.* considered the enzymatic breakdown of fatty acids of particular importance (Cramer *et al.* 2005). This was because barley contains lipoxygenase and other enzymes responsible for the oxidation of fatty acids, leading to the production of the compounds just mentioned. Indeed, barley varieties have been developed lacking the lipoxygenase enzyme involved in lipid breakdown in order to reduce the level of (E)-2-nonenal in beer (Yu *et al.* 2014). The same sort of enzymatic fatty acid oxidation also gives the characteristic flavors of cucumbers, mushrooms, tomatoes, peas, and beans. It is the differences in the fatty acids and the lipoxygenases present that provides a different set of flavor compounds in each (Moir 1992).

In Scotch Grain Whisky production, other cereals are additionally used, most commonly wheat and corn. Occasionally other cereals such as rye may be used to produce Scotch Grain Whisky. In other whisky regions such as the United States and Canada, rye is a more common ingredient. From the limited literature, it seems likely that the compounds contributing to the aromas of these cereals will be similar to that of barley but the proportions will be different (Maga 1978).

While at least some of the barley derived compounds are found in the final whisky, the intervening processing steps have a large impact on the final levels. As much as providing a small number of flavor compounds directly, the cereal also provides the raw material for the formation of flavor compounds during the production process. There is therefore potential for variation between barley variety and barley growing conditions to contribute to the formation of flavor compounds. However, the narrowing of the genetic base of current barley varieties means that there is limited potential for more modern varieties to produce varied flavor characteristics. Significant differences tend to be seen in older barley varieties, but these come at a cost of reduced alcohol yield. Growing conditions are known to impact on barley composition, for example by influencing total nitrogen content, and this in turn can influence flavor compound production in the process (Hill and Stewart 2019; Kyraleou *et al.* 2021).

4.4.2 Malting

The first step in the malt whisky production process is to malt the barley. The main purpose of this process is to modify the barley grain to ensure that starch is accessible to endogenous amylolytic enzymes later in the process. In doing so, the aroma profile derived from the raw barley is developed. Tressl *et al.* suggest three routes for flavor compound formation during malting (Tressl, Bahri, and Helak 1983):

- The enzymatic and chemical oxidation of unsaturated fatty acids
- The Maillard reaction of free amino acids and reducing sugars
- The thermal degradation of biosynthesized precursors like S-methyl methionine and cinnamic acid derivatives

Based on this, a key stage for flavor development is the application of heat to the malted barley during kilning. The kilning process not only results in some of the barley compounds increasing in level but also a decrease in the level of others (Dong *et al.* 2013). The grassy aldehydes hexanal and (E)-2-hexenal have been observed to decrease during kilning, while the malty aldehydes 2-methyl butanal and 3-methylbutanal, along with the stale aldehyde (E)-2-nonenal have been observed to increase in level. 2-Methylbutanal and 3-methylbutanal are formed from isoleucine and leucine via the Strecker degradation part of the Maillard reaction. (E)-2-nonenal is formed from the thermal breakdown of unsaturated fatty acids, notably linoleic acid (Belitz, Grosch, and Schieberle

2009). In line with these changes, there is a reduction in green aroma and an increase in malty aroma moving from barley to malt.

Not only are barley aroma compounds increased or decreased during malting, but new compounds are also formed. During germination, S-methylmethionine, an intermediate in many biosynthetic pathways, is formed by enzymatic reaction (Tressl, Bahri, and Helak 1983; Ferreira and Guido 2018). During kilning this labile precursor is decomposed into dimethyl sulfide. Dimethyl sulfide has a vegetable-like, sweetcorn flavor. Another important flavor compound formed during kilning is methional (Piornos *et al.* 2020). This compound has a boiled potato-like aroma and, analogous to 2-methyl butanal and 3-methylbutanal, is formed from the amino acid methionine via the Strecker degradation. Not only is methional a highly aroma-active compound itself, it also acts as a precursor for other potent aroma compounds which will be described later in the process (Furasawa 1996).

These compounds are formed during the production of typical pale distillers malt, which is kilned at a relatively low temperature to retain enzyme activity. Occasionally, to provide different flavors, distillers have used colored malt that has been kilned to a higher temperature. As the kilning temperature is increased, other groups of compounds start to play a significant role in the aroma of the malt. At high temperatures, the thermal breakdown of phenolic precursors such as ferulic acid can produce flavor-active compounds such as vinylguaiacol, ethylguaiacol, guaiacol, and 4-methylphenol (Tressl, Bahri, and Helak 1983; Lentz 2018; Scholtes, Nizet, and Collin 2014). These compounds can give spicy, smoky, and phenolic aromas (The Good Scents Company n.d.). At such higher temperatures, nitrogen compounds are also produced via Maillard reactions. Reaction of reductones with amino acids gives alpha-amino ketones, which condense together to form pyrazines (Moir 1992). Relative to other Maillard reaction flavor compounds, the pyrazines require a high level of thermal energy to form, so they are formed at relatively high temperature and low moisture content (Coghe *et al.* 2012). Pyrazines found in roasted malt include pyrazine, methylpyrazine, 2,3-dimethylpyrazine, 2,6-dimethylpyrazine, ethylpyrazine, and 2-ethyl-3-methylpyrazine (Coghe *et al.* 2012). These compounds have aromas variously described as nutty, roasted, green, earthy, and fruity (Moir 1992).

In Scotch Malt Whisky production, the application of peat smoke to the drying malt during kilning is sometimes used as a source of flavor. Peat would have been used as a heat source in the kiln in the past, but now more efficient fuels have replaced it. However, certain Scotch Whisky producers continue to apply peat smoke to the malt, no longer to dry the malt but now solely to provide a particular flavor. Important flavor compounds created by the partial combustion of peat include guaiacol, 4-methylguaiacol, 4-ethylguaiacol, 2-methyl phenol, 3-methylphenol, 4-methylphenol, and 4-ethylphenol (Harrison 2007).

These compounds are formed from partially decomposed lignin and lignin-like material in the peat. Some of the compounds, notably the guaiacols, produced from peat combustion can also be observed in roasted malt (Tressl, Bahri and Helak 1983; Lentz 2018; Scholtes, Nizet, and Collin 2014). A key difference in the aroma of peat smoke compared to roasting the malt is the presence of medicinal aroma in the peat smoke. This likely occurs, at least in part, due to a greater abundance of the alkyl phenols in peat smoke (Harrison 2007). These compounds tend to have more of a medicinal aroma, whereas the guaiacols tend to have more of a smoky aroma. The relative abundance of the alkyl phenols may in turn relate to an abundance of sphagnum moss in peat (Harrison 2007). Sphagnum does not contain lignin but an analogous material lacking the methoxy groups associated with true lignin which identify the guaiacols. The partial decomposition of peat as it ages may also result in the removal of methoxy groups contributing to a higher proportion of alkyl phenols. Peats from different sources have been shown to contain different profiles of guaiacols and alkyl phenols and therefore give different aroma profiles (Harrison 2007). This difference in peat composition was considered to be a result of differences in vegetation input into the peat and the level of decomposition that vegetation has undergone.

4.4.3 Mashing

Research into the impact of mashing on flavor has largely focused on the relative turbidity of the resulting wort and on the effect of the long-chain unsaturated fatty acids associated with increasing turbidity on flavor compound development during the subsequent fermentation (Bathgate 2019; Kühbeck, Back, and Krottenthaler 2006). In both malt and wort, linoleic acid (C18:2) and palmitic acid (C16) are the most abundant long-chain fatty acids, while oleic acid (C18:1), linolenic acid (C18:3), and stearic acid (C18) are present at lower levels (Anness and Reed 1985a). The unsaturated fatty acids have the greatest influence on flavor compound development given the increased rate of oxidation associated with each additional double bond (Belitz, Grosch, and Schieberle 2009).

Flavor compounds from unsaturated fatty acids of particular note, based on their low aroma detection thresholds, include (E)-2-nonenal, (E,E)-2,4-decadienal, (E,Z)-2,6-nonadienal, and 1-octen-3-one (Boothroyd 2013; Wanikawa, Hosoi, and Kato et al. 2002). The aromas of these compounds are diverse: (E)-2-nonenal has been described above as having a stale aroma, (E,E)-2,4-decadienal may be classed as fatty, (E,Z)-2,6-nonadienal as green, and 1-octen-3-one as earthy (The Good Scents Company n.d.). Of these four compounds, (E)-2-nonenal, (E,E)-2,4-decadienal, and 1-octen-3-one are produced from linoleic acid oxidation, while (E,Z)-2,6-nonadienal is produced from linolenic acid (Belitz, Grosch, and Schieberle 2009).

The extraction of long-chain fatty acids during mashing is inefficient, and the large majority are left behind in the mash tun, retained in spent grains (Anness and Reed 1985a). Factors influencing their extraction are therefore of great importance in terms of flavor development. The type of vessel used for mashing can play a role in the level of wort turbidity and associated long-chain fatty acids. Three types of mashing vessel are used in Scotch Malt Whisky production: modern (semi-)lauter tuns, traditional mash tuns, and mash filters. Lauter tuns, which are now the most common mashing vessel in Scotch Malt distilleries, filter faster than traditional mash tuns but tend to yield a more turbid wort (Stewart, Yonesawa, and Martin 2007). The mash filter, used only in a very small number of Scotch Malt distilleries, can filter the mash quickly but may yield a more turbid wort than the lauter tun (Kühbeck, Back, and Krottenthaler 2006; Anness and Reed 1985b). It is noted though, that within each mashing vessel type, process changes can be made to adjust wort turbidity. For example, the increase in turbidity from a mash filter is largely related to high levels in the initial runoff, and wort recirculation is one approach that could be used to diminish that effect (Anness 1985a; Golston 2021). When raking is used in mash and lauter tuns, a correlation exists between the intensity of raking, and therefore filtration speed, and wort turbidity (Kühbeck, Back, and Krottenthaler 2006). Sparging—that is, adding increasingly hot water to the mash to extract residual extract—can also increase the level of unsaturated fatty acids associated with increased turbidity. Experiments comparing sparge temperature showed that increasing the temperature increases turbidity and unsaturated fatty acids in the wort (Stewart, Yonesawa, and Martin 2007).

While much work has been done looking at the impact of wort unsaturated fatty acids on flavor compound production during fermentation, there is also the potential for unsaturated fatty acids to act as flavor precursors during the mashing process. Lipase acts to release unsaturated fatty acids from lipids, and this enzyme is thought to be thermally stable during mashing (Kobayashi et al. 1994). The unsaturated fatty acids are oxidized into their corresponding hydroperoxides either through lipoxygenase-catalyzed oxidation or heat-induced auto-oxidation (Kobayashi et al. 1994). The relative importance of each pathway is unclear, but it is known that lipoxygenase is less thermally stable than lipase at mashing temperatures, and so its influence will be reduced as temperature is increased (Kobayashi et al. 1993; Schwarz, Stanley, and Solberg 2002). The resulting free fatty acid hydroperoxides are unstable and may degrade during mashing to create the carbonyl flavor compounds previously mentioned. Additionally, free unsaturated fatty acids or their hydroperoxides surviving mashing may break down to form flavor compounds later in the production process (Stewart, Yonesawa, and Martin 2007).

The composition of water added into the process at the mashing stage has in the past been assumed to have a highly significant impact on flavor (Hastie

1926). More recent research has shown that, while mashing water composition can impact on flavor, the effect is relatively subtle with the level of organic material in the water having the most significant impact (C. A. Wilson 2008). The flavor chemistry underlying the effect of water on flavor is still not clear.

4.4.4 Fermentation

Fermentation is a key point in the production process for the formation of flavor compounds. Yeast utilizes the wort nutrients to produce not only ethanol but also an array of secondary metabolites that contribute to flavor. Quantitatively, the main groups of flavor compounds produced by yeast are alcohols and esters (Salo, Nykanen, and Suomalainen 1972). Though less abundant, acids, carbonyls, and sulfur compounds are also produced by yeast and make a contribution to flavor (Salo, Nykanen, and Suomalainen 1972; Pires *et al.* 2014). The compound profile at the end of fermentation is dependent on a combination of factors including wort composition, yeast strain, and fermentation variables such as time and temperature.

4.4.4.1 Esters

Esters are a large group of compounds produced by yeast during fermentation and generally considered to contribute desirable flavors. The esters can be split into two groups depending on their formation pathway: the ethyl esters of fatty acids and the acetate esters.

Important fatty acid ethyl esters contributing to Scotch Malt Whisky flavor are the medium-chain fatty acid (MCFA) ethyl esters, including ethyl hexanoate, ethyl octanoate, ethyl decanoate, and ethyl dodecanoate (Salo, Nykanen, and Suomalainen 1972). Of these, shorter-chained ethyl esters such as ethyl hexanoate have more of a fresh fruit aroma while, as the chain gets longer, the aroma tends to become more waxy (The Good Scents Company n.d.). Though there is some debate as to why they are formed, the fatty acid ethyl esters are formed as by-products of fatty acid synthesis by yeast. Pyruvate produced from the glycolysis of fermentable sugars is converted to acetyl CoA. Acetyl CoA is then used to produce the various fatty acid CoAs that, when combined with ethanol, form the ethyl esters (Walker and Hill 2016). Saerens *et al.* showed that the formation of the majority of the MCFA ethyl esters in yeast is catalyzed by two acyl-CoAs: ethanol O-acyltransferases (AEATases), Eeb1 and Eht1 (Saerens *et al.* 2006).

Important acetate esters contributing to Scotch Whisky flavor are ethyl acetate, isoamyl acetate, and phenylethyl acetate (Salo, Nykanen, and Suomalainen 1972). Ethyl acetate has an ethereal aroma, isoamyl acetate a fruity aroma, and phenylethyl acetate a floral aroma (The Good Scents Company n.d.). Acetate esters are formed when ethanol, or higher alcohols produced from amino acids via the Ehrlich pathway, are combined with acetyl CoA (Walker and Hill 2016). The alcohol acetyl transferases I and II (AATase I and II), encoded by the genes ATF1 and ATF2, catalyze the synthesis of acetate esters (Saerens *et al.* 2010).

The level of ester formation is dependent on the concentrations of the substrates and the activity of the enzymes involved in their synthesis and hydrolysis (Saerens *et al.* 2018). In the case of the acetate esters, the expression level of ATF1 and ATF2 have been considered the most important factor determining their levels during fermentation (Verstrepen *et al.* 2003). In the case of the ethyl esters, rather than the level of enzyme expression, it is the precursor concentrations that have been considered as the most important limiting factor for their synthesis (Saerens *et al.* 2010, 2018).

Many factors can influence ester production, but wort components that are well-known for their function as negative regulators of ester formation are long-chain unsaturated fatty acids, notably linoleic acid, and oxygen (Fujiwara *et al.* 1998; Taylor, Thurston, and Kirsop 1979; Anderson and Kirsop 1975; Saerens *et al.* 2010, 2018; Stewart, Yonesawa, and Martin 2007). These components are known to stimulate yeast growth and so enable efficient fermentation, but in doing so they limit ester formation (Taylor, Thurston, and Kirsop 1979). In this way, the unsaturated fatty acids associated with turbid wort not only contribute flavor compounds via their degradation but also influence the production of esters during the fermentation process.

Other wort components that can contribute to the ester profile include the total sugar content of the wort, reflected in the wort specific gravity, and the free amino nitrogen (FAN) content. A relative increase in acetate esters as wort gravity is increased is well documented in brewing literature and is a consideration for distillers as they move to higher gravities to improve efficiency (Anderson and Kirsop 1974; Saerens *et al.* 2018). Similarly, it has also been shown that a higher level of FAN can increase acetate ester production but not ethyl ester production (Saerens *et al.* 2018). In this way, variation in wort composition parameters can be used to control the final ester profile.

Fermentation parameters such as temperature and time may play a role in ester formation. It is uncommon for malt whisky fermentation vessels to be temperature controlled, though as wort gravities increase, this may become required to limit the temperature rise and stress on the yeast as the fermentation progresses. Currently, in most cases the temperature profile of the fermentation is dictated by the temperature of the wort when the yeast is added, in combination with ambient conditions during the process. Possibly because control has been limited, there is little literature on the influence of temperature on malt whisky fermentations. From the limited work where the temperature of fermentation has been investigated, it has been suggested that higher temperatures result in lower levels of esters (Ramsay and Berry 1983). This will be a consideration where temperature control is applied to higher gravity fermentations.

The acetate esters and MCFA esters are primarily formed during the active phase of yeast fermentation, and so their levels increase over this period (Stewart 2017). As that period ends, their levels will start to plateau or may even decrease in some cases due to evaporation or possibly due to the influence of hydrolyzing enzymes found in the yeast (Suomalainen 1981). Lactic acid bacteria, active in a secondary fermentation as the yeast starts to die off, have then been observed to increase ester levels (Barbour and Priest 1988; Geddes and Riffkin 1989). This effect may be either direct by an ester-forming function of the bacteria or indirect by the bacteria producing precursors such as acetic and lactic acids, which may form esters, predominantly with ethanol, later in the process. Longer fermentations are often associated with a more fruity new make spirit, and this secondary fermentation by lactic acid bacteria may be a contributing factor. Differences in the strain or strains of lactic acid bacteria present is likely to result in variation in the impact of secondary fermentation (N. R. Wilson 2008).

Yeast strain is another factor that is known to contribute to the acetate and ethyl ester profile created during fermentation (Stewart 2017). Currently in Scotch Malt Whisky production, the majority of producers use relatively few strains of *S. cerevisiae* (Walker and Hill 2016). In theory, though, a large selection of potential yeast strains can be exploited, and this is an area that is being explored more by various producers.

A third group of esters—lactones or cyclic carboxylic esters—can also be formed during fermentation. γ-decalactone and γ-dodecalactone have previously been detected in Scotch Malt Whisky and confer sweet, fatty flavor (Wanikawa, Hosoi, and Takise *et al.* 2012; Wanikawa, Hosoi, and Kato 2000; Wanikawa, Hosoi, and Shoji *et al.* 2001; Wanikawa, Shoji *et al.* 2002). It has been shown that lactic acid bacteria can convert (9Z)-hexadecenoic and (9Z)-octadecenoic acid to 10-hydroxyhexadecanoic and 10-hydroxyoctadecanoic acid, respectively. The precursor hydroxy fatty acids are thought to then be converted to the lactones by distiller's yeast. The amounts of hydroxy fatty acids have been found to be increased when brewer's yeast was used together with distilling yeast, owing to the promotion of lactic acid bacteria growth by the death of brewer's yeast at an early stage of fermentation (Wanikawa, Hosoi, and Kato 2000). It is possible that wild yeast naturally present in the fermentation could potentially contribute a similar effect to brewer's yeast (Neri 2006; N. R. Wilson 2008). It is notable that in Scotland the use of brewer's yeast was largely phased out by distillers during the late 1990s and mid-2000s (Walker and Hill 2016). Although the benefits of using brewer's yeast for whisky fermentations have been reported, the practicalities of using it have meant that brewer's yeast is used by only a very small number of distilleries now (Dolan 1976).

4.4.4.2 Carbonyls

Key aroma carbonyls that have been mentioned previously include the Strecker aldehydes such as 2-methylpropanal, 2-methylbutanal, 3-methylbutanal, and methional; and the lipid-derived compounds such as (E)-2-nonenal, (E,E)-2,4-decadienal, and (E,Z)-2,6-nonadienal. During fermentation these compounds, characteristic of the wort, are reduced in level by the action of the yeast (Kłosowski *et al.* 2017; Peppard and Halsey 1981; Perpète and Collin 1999). (E)-2-nonenal has been found to be reduced by yeast to the alcohol 1-nonanol, which in turn was esterified to yield nonyl acetate (Peppard and Halsey 1981). A similar fate might be expected for the other carbonyls mentioned here (Smit, Engels, and Smit 2009). The reduction in carbonyl levels over the course of fermentation combines with the increase in ester levels during the process to help explain the transition from heavier aromas such as oily, cereal, and nutty from shorter fermentations to lighter aromas such as fruity in longer fermentations.

4.4.4.3 Higher Alcohols

Higher alcohols such as 2-methylpropannol, 2-methylbutanol, and 3-methyl butanol are known to play a role, albeit a relatively limited one, in the aroma of Scotch Whisky (Salo, Nykanen, and Suomalainen 1972). They are thought to contribute to a general intensification of alcoholic or solvent-like aroma, with 3-methylbutanol also reported to contribute to fruity and sweet aromas (Van Laere *et al.* 2008). Two important pathways lead to the production of higher alcohols in fermentation: the catabolic route (Ehrlich pathway) and the anabolic route (Genevois pathway). The Ehrlich pathway has been identified in yeasts as the main route for higher alcohol formation, particularly in early fermentation when free amino acids are plentiful (Smit, Engels, and Smit 2009). For aromatic and branched-chain amino acids, the α-ketoacids resulting from their transamination on this pathway are not intermediates of central metabolism and so are often transformed by the yeast cells before being excreted into the growth medium (Vuralhan, Luttik *et al.* 2005; Hazelwood *et al.* 2006). On this pathway, Strecker aldehydes are formed from the α-ketoacids (Smit, Engels, and Smit 2009). Therefore, while these aldehydes are generally reduced in level over the course of fermentation through reduction of those present in wort to the corresponding alcohols, their formation via the Ehrlich pathway may prevent complete removal.

4.4.4.4 Acids

Also produced via the catabolism of amino acids by yeast are carboxylic acids. Acids produced in this way that are of potential interest from a flavor perspective are 2-methylbutanoic acid and 3-methylbutanoic acid (Salo, Nykanen, and

Suomalainen 1972). Depending on the redox status of the cell, aldehydes can be reduced by alcohol dehydrogenases or oxidized to the corresponding carboxylic acid by aldehyde dehydrogenases (Hazelwood *et al.* 2006; Vuralhan *et al.* 2003). In this way, a higher level of oxygen availability has been found to increase the level of the carboxylic acids (Vuralhan *et al.* 2003). Though the carboxylic acids 2-methylbutanoic acid and 3-methylbutanoic acid themselves are considered to have a low impact on whisky aroma, the esters they form with ethanol have lower aroma detection thresholds and may play a greater role (Ferreira, Lopez, and Cacho 2000). Ethyl (S)-2-methylbutanoate and ethyl 3-methylbutanoate have fruity aromas and have been reported as having a high aroma impact in bourbon, and it is possible that this will also be the case for Scotch Whisky (Poisson and Schieberle 2008).

Acetic acid is another acid formed during fermentation that can have an impact on whisky aroma. Its sour aroma is well-known, but that is more likely to be detected as a defect than a typical Scotch Whisky aroma. A more important function for acetic acid in fermentation is to act as a precursor to other aroma compounds such as the acetate esters. If Acetyl-CoA is not required as a biosynthetic building block, the CoA is removed, and the acid is either lost to the fermentation medium or enzymatically stabilized by esterification (Campbell 2003). As discussed, another potential source of acetic acid in whisky fermentation is lactic acid bacteria.

4.4.4.5 Sulfur Compounds

Some sulfur compound formation takes place during fermentation, but it is also important as a point where other sulfur compounds are removed. For example, the concentration of dimethyl sulfide present after fermentation is down to a balance of formation and removal. As a highly volatile compound, some of the dimethyl sulfide found in wort is lost through evaporation (Anness and Bamforth 1982). However, some formation will also take place, as dimethyl sulfoxide in wort may be reduced by yeast to dimethyl sulfide during fermentation, so removal may not be complete (Ferreira and Guido 2018).

During fermentation, two pathways important for the production of sulfur flavor compounds are the biosynthesis of sulfur-containing amino acids and the reduction of sulfate salts present in wort (Stewart 2017; Campbell 2003). The formation of methional is analogous to other Strecker aldehydes, wherein it is formed as an intermediate in the Ehrlich pathway as methionine is converted to methionol (Landaud, Helinck, and Bonnarme 2008). Factors controlling the formation of methional will be similar to that of other Strecker aldehydes. Further enzymatic or chemical breakdown of methional can yield methanethiol, an important precursor to sulfur compounds formed during distillation (Furasawa 1996; Landaud, Helinck, and Bonnarme 2008).

Though hydrogen sulfide is known to be an aroma compound formed during fermentation, its volatility means that it is not relevant to mature whisky (Quain 1989). However, it is also thought to be an important precursor for the production of dimethyl trisulfide during distillation (Furasawa 1996). One way hydrogen sulfide can be formed by yeast is as an intermediate in the utilization of sulphate in the biosynthesis of the sulfur-containing amino acids methionine and cysteine (Quain 1989). It is thought that hydrogen sulfide levels will increase toward the end of fermentation if amino acid levels in the wort drop and biosynthesis of the sulfur-containing amino acids can no longer continue, but sulphate is still available and so hydrogen sulfide continues to be formed as an intermediate (Landaud, Helinck, and Bonnarme 2008; Doyle and Slaughter 1998).

4.4.5 Distillation

A key function of distillation is to physically separate compounds, formed during preceding process steps, that are desirable for the final product from those that are not. The various factors contributing to the physical separation of compounds during distillation will not be covered here. In carrying out the distillation process, the application of heat, as well as the presence of copper, present opportunities for the formation of flavor compounds.

4.4.5.1 Sulfur Compounds

The use of copper to construct pot stills is well-known for its ability to remove undesirable sulfur compounds from the distillate (Harrison *et al.* 2011). However, distillation is also a point at which sulfur compounds are formed. Sulfur compounds of particular note formed during distillation are dimethyl trisulfide and methyl-2-methyl-3-furyl disulfide (Watts 2005; Furasawa 1996; Harrison *et al.* 2011). The aroma of these two sulfur compounds are quite distinct, with dimethyl trisulfide described as cooked vegetables and methyl-2-methyl-3-furyl disulfide described as meaty (Furasawa 1996; The Good Scents Company, n.d.; Watts 2005).

Methanethiol is a key precursor for the sulfur compound formation during distillation (Watts 2005; Furasawa 1996). Methanethiol is highly aroma-active itself but unlikely to be of relevance post-maturation given its volatility. It may be formed during fermentation, as previously described, or during distillation from the breakdown of methionine and subsequently methional (Landaud, Helinck, and Bonnarme 2008; Watts 2005; Furasawa 1996). Thiamine breakdown, via a number of intermediates, provides an additional precursor that reacts with methanethiol to give methyl-2-methyl-3-furyl disulfide (Watts 2005). Hydrogen sulfide provides the additional precursor to form dimethyl trisulfide (Furasawa 1996).

Interestingly, copper in the wash still pot may contribute to elevated levels of dimethyl trisulfide and methyl-2-methyl-3-furyl disulfide precursors promoting formation (Furasawa 1996; Watts 2005). For example, it is thought that copper ions can promote the breakdown of methionine to methional (Furasawa 1996). However, copper present later in the distillation process will reduce the levels of dimethyl trisulfide and methyl-2-methyl-3-furyl disulfide, with the overall effect that the use of copper in stills will result in a reduction in sulfur-like aromas in new make spirit (Harrison *et al.* 2011). Sometimes it is desirable to retain an elevated level of sulfur compounds in new make spirit to provide a heavier aroma, but mostly it is desirable to limit sulfur compound levels. This can be achieved by increasing copper contact through increasing reflux, increasing the temperature of cooling water in the condensers, as well as increasing the amount of rest that the copper has had between distillations. Also, decreasing the pH of the still charge may also increase sulfur compound removal by increasing the availability of fresh copper surface associated with sulfur compound removal (Thulasidas 2007).

4.4.5.2 Carbonyl Compounds

After the fermentation process has acted to reduce the levels of flavor-active Strecker aldehydes and lipid-derived carbonyls, the heat input during distillation provides a further opportunity for the production of these groups of compounds from precursors not utilized by yeast during fermentation.

In addition to previously mentioned aldehydes such as 2-methylpropanal, 2-methylbutanal, 3-methylbutanal and methional, another compound formed via the Maillard reaction of potential importance to flavor is 2-acetyl-1-pyrroline (Conner, Jack, and Walker 2010). This compound has been found in both new make spirit and matured whisky and is highly aroma-active with an aroma detection threshold of 67 ppt in 20% ethanol and a cereal aroma (Conner, Jack, and Walker 2010). 2-Acetyl-1-pyrroline is formed from proline, which tends to be the most abundant amino acid in the wash at the end of fermentation (Freeman *et al.* 1999; Wei *et al.* 2017; Schieberle 1989). Therefore, of the compounds formed via the reaction between amino acid and sugar, 2-acetyl-1-pyrroline formation might be considered of particular relevance during distillation.

Changes in distilling process over time may have contributed to a lessening of the importance of Maillard reactions during distillation. Although the Maillard reaction can happen at any temperature, increasing the temperature will increase the rate and extent of the reaction until an optimum temperature is reached, beyond which caramelization and pyrolysis reactions will occur. A shift from directly heating stills by anthracite or coke furnaces to indirect heating by steam coils or external calandria has caused a reduction in heating temperature during distillation and the potential for Maillard reactions (Bathgate 2019). In 1928, Hastie considered furfural, another Maillard reaction product, as one of the important substances contributing to the character of whisky (Hastie and Dick 1928). However, the levels detected in modern-day new make distillates suggest it is no longer of such significance to flavor. It is likely that other Maillard reaction products are also of lesser significance now.

Heating during distillation will break down the unsaturated fatty acids discussed earlier, such as linoleic acid, as well as their ethyl esters formed during fermentation. This breakdown of unsaturated fatty acids yields important aroma-active carbonyls such as (E)-(E)-2-nonenal, (E,E)-2,4-decadienal, (E,Z)-2,6-nonadienal, and 1-octen-3-one (Boothroyd 2013). It is likely that the production of these compounds has been less influenced by reducing distillation temperatures than that of the Maillard reaction products, given the labile nature of the unsaturated fatty acids as well as their intermediates (Su 2003). Also, another interesting potential is for unsaturated fatty acids and their ethyl esters to break down during storage in receiver vessels used to collect recycled fractions of foreshots and feints, particularly if deposits are allowed to form with copper ions present as a pro-oxidant (Lukić *et al.* 2011; Wasowicz *et al.* 2004).

Components from both the lipid oxidation pathway and the Maillard reaction can also interact to influence the flavor compound profile. 2-Pentylfuran is a flavor compound reported in new make spirit and has a buttery, green bean aroma (Daute *et al.* 2021; Krishnamurthy *et al.* 1967; Belitz, Grosch, and Schieberle 2009). 2-Pentylfuran is formed from linoleic acid via (E)-2-nonenal, but the conversion of (E)-2-nonenal to 2-pentylfuran has been found to be enhanced by the presence of amino acids (Adams *et al.* 2011). In this way, the different reaction pathways occurring during distillation interact, and so the overall composition of the wash as the starting substrate needs to be considered to understand what will be formed during the process.

The carbonyl compound beta-damascenone has been found to be formed during distillation (Masuda and Nishimura 1980). This floral compound is notable for having a low aroma detection threshold but also a low intensity index, meaning that its impact is lower than its aroma detection threshold might suggest (Perry 1983; Lee *et al.* 2001). It has been reported that beta-damascenone is formed from precursors found in malt (Masuda and Nishimura 1980). Though these precursors have not been identified, analogous to other plant materials, it may be that a glycoside precursor is hydrolyzed and then undergoes further acid-catalyzed rearrangements to produce beta-damascenone (Hjelmeland and Ebeler 2015). Indeed, the work of Masuda and Nishimura showed that beta-damascenone formation during distillation was enhanced under acidic conditions (Masuda and Nishimura 1980). Therefore, a more acidic wash would be expected to lead to greater production of this compound.

4.4.5.3 Esters

Esterification of alcohols and carboxylic acids is possible during distillation, with the low pH and presence of copper ions potentially contributing to the process (Watson 1983a). However, the significance of this reaction is thought to be small in comparison to the formation of esters during fermentation.

4.4.6 Maturation

During maturation, a number of chemical and physical changes take place that are important for flavor development. The cask is not airtight, and through evaporation there will be a loss of volatile flavor compounds with a boiling point lower than ethanol (Hasuo and Toshizawa 1986). Those flavor compounds with a boiling point higher than ethanol will remain in the cask and may actually increase in concentration due to ethanol evaporation. The inner surface of the cask can remove compounds. The inner surface of ex-bourbon casks are charred and the inner surface of ex-sherry casks are toasted. A charred surface of the cask is known to remove certain compounds, notably flavor-active sulfur compounds, and so there will be some loss from the maturing spirit via that route (Philp 1986; Fujii, Kurokawa, and Saita 1992). Two key areas for flavor compound formation are the decomposition and extraction of wood compounds; and those involving the chemical reaction of new make spirit compounds with one another and with wood compounds. The extent of these reactions is influenced by a number of factors, but cask type and maturation time are considered the most important, with other factors such as fill strength and warehouse conditions also likely to play a role (Conner 2014).

4.4.6.1 Extraction of Wood Compounds

Maturation has a significant impact on the complexity of the whisky, with the number of compounds detected moving from hundreds in new make spirit to thousands in mature whisky (Kew 2016; Roullier-Gall *et al.* 2018). However, only a small number of wood-derived compounds are actually known to have a direct impact on flavor (Conner 2014; Poisson and Schieberle 2008).

Wood can be described as a three-dimensional biopolymer composite composed of an interconnected network of cellulose, hemicelluloses, and lignin with minor amounts of extractives and inorganics (Rowell *et al.* 2012). From a flavor point of view, in Scotch Whisky maturation, the extractives and lignin might be considered of greatest importance. Due to its crystalline structure, cellulose undergoes only a few chemical degradations or modifications (Le Floch, Jourdes, and Teissedre 2015). Heating hemicellulose can yield flavor compounds such as furfural, 5-methyl-2-furaldehyde, cyclotene, and maltol (Le Floch, Jourdes, and

Teissedre 2015). However, these compounds are not found at high enough levels in Scotch Whisky to have a direct impact on flavor (Conner 2014).

Of compounds derived from lignin, vanillin is a key contributor to flavor in Scotch Whisky, with a well-known vanilla aroma. It is present naturally in oak wood, either free or glycosidically bound, but the heat treatment applied to casks during manufacture greatly increases the level of vanillin through the thermal degradation of lignin (Nishimura *et al.* 1983; Bloem, Lonvaud-Funel, and de Revel 2008; Slaghenaufi *et al.* 2016). Nishimura *et al.* showed that heating wood up to approximately 200°C increased the level of vanillin, but increasing further to charring temperature decreased the level through volatilization and carbonation (Nishimura *et al.* 1983). However, in a cask, the formation of char on the surface disrupts wood structure and allows spirit to penetrate further into the wood and access thermally degraded lignin underneath (Perry, Ford, and Burke 1990).

Other phenolic compounds derived from wood and thought to contribute to flavor include eugenol, guaiacol, and ethylguaiacol (Conner 2014). Whereas vanillin contributes vanilla aroma, these other phenolics are thought to contribute spicy aromas (The Good Scents Company, n.d.). Eugenol, guaiacol, and ethylguaiacol are known to be formed by thermal breakdown of lignin (Brebu and Vasile 2010). However, in the case of eugenol, the level released from lignin is considered of less importance than the levels of free and glycosidically bound compound found in wood (Conner 2014).

Another key aroma compound found free and glycosidically bound in oak wood is oak lactone (3-methyl-4-octanolide). Theoretically, there are four stereoisomers of 3-methyl-4-octanolide, but only two isomers have been found in nature; namely, the (3S, 4S) and (3S, 4R) isomers or the respective *cis-* and *trans*-oak lactones (Masson *et al.* 2000). The two isomers both have coconut aromas; however, it has been reported that there are differences in the nature of that coconut aroma between the two (Noguchi 2016). Heat treatment is not considered an important factor contributing to the levels of these compounds, and so the levels available to extract into the spirit are dictated by the content of the untreated wood (Noguchi 2016). The relative levels of the two isomers vary between oak species and even between forest regions (Masson *et al.* 2000; Noguchi 2016; Waterhouse and Towey 1994). Of oak species used for Scotch Whisky production, American oak has been found to have the cis isomer dominating, while European oak has an equal ratio. Outside Scotch Whisky, Japanese oak has been found to contain the *trans*-isomer predominantly.

4.4.6.2 Reaction of New Make Spirit Compounds

During maturation, some new make spirit compounds can react to form different compounds and alter the flavor of the spirit. Alcohols may be oxidized to aldehydes and acids, and, in turn, alcohols can also react with aldehydes to form acetals and with acids to form esters (Reazin 1981). A shift from alcohols, aldehydes, and acids to acetals and esters is associated with a shift from heavier aromas such as cereal, oily, and sour to lighter aromas such as fruity and solventy.

Oxidation rate is thought to be influenced by a combination of hydrolysable tannins, dissolved oxygen and copper ions, which contribute to the production of active oxygen (Philp 1986). Therefore, factors contributing to the levels of these components in the spirit will contribute to the level of oxidation. For example, copper levels at the start of maturation will vary according to the level of extraction from the copper stills during distillation. During maturation, the level of copper will decrease due to adsorption to wood, and it has been suggested that this adsorption is greatest when the wood is charred rather than toasted (Muller and McEwan 1998).

Both acetal formation and ester formation are equilibrium reactions, so various factors may influence the relative rates of the forward and reverse reactions. For example, the equilibrium between free aldehyde, hemiacetal intermediate, and acetal has been shown to be affected by spirit pH and ethanol concentration (Perry 1986). During maturation, the concentration of ethyl esters increases due to the esterification of free acids by ethanol. Le Chatelier's principle plays a role here with an excess of the ethanol precursor favoring the forward reaction of ester formation. Transesterification reactions are also thought to occur, which in the presence of the large excess of ethanol again favors the formation of ethyl esters. So isoamyl acetate, for example, is likely to be converted to ethyl acetate.

4.5 SUMMARY

Despite the strict legal definition governing the Scotch Malt Whisky production process, a large amount of flavor diversity is still found across different brands. This diversity is created by variation in a number of factors at all the steps of the production process. From the information gathered in this chapter, fermentation and maturation might be considered of particular importance for flavor compound formation. However, the intervening process steps are just as critical, given that what goes in dictates what comes out. For example, variations in mashing process will influence wort composition and therefore the flavor compounds formed during fermentation. The subsequent distillation step then adds further variation to the composition of the new make spirit placed in a cask for maturation.

REFERENCES

Adams, A., C. Bouckaert, F. Van Lancker, B. De Meulenaer, and N. De Kimpe. 2011. Amino acid catalysis of 2-alkylfuran formation from lipid oxidation-derived α,β-unsaturated aldehydes. *Journal of Agricultural and Food Chemistry* 59:11058–11062.

Anderson, R. G., and B. H. Kirsop. 1974. The control of volatile ester synthesis during the fermentation of wort of high specific gravity. *Journal of the Institute of Brewing* 80:48–55.

Anderson, R. G., and B. H. Kirsop. 1975. Oxygen as a regulator of ester accumulation during the fermentation of wort of high specific gravity. *Journal of the Institute of Brewing* 81:111–115.

Anness, B. J., and C. W. Bamforth. 1982. Dimethyl sulfide—a review. *Journal of the Institute of Brewing* 88:244–252.

Anness, B. J., and R. J. R. Reed. 1985a. Lipids in the brewery—a material balance. *Journal of the Institute of Brewing* 91:82–87.

Anness, B. J., and R. J. R. Reed. 1985b. Lipids in wort. *Journal of the Institute of Brewing* 91:313–317.

Barbour, E. A., and F. G. Priest. 1988. Some effects of Lactobacillus contamination in scotch whisky fermentations. *Journal of the Institute of Brewing* 94:89–92.

Bathgate, G. N. 2019. The influence of malt and wort processing on spirit character: The lost styles of Scotch malt whisky. *Journal of the Institute of Brewing* 125:200–213.

Beal, A. D., and D. S. Mottram. 1994. Compounds contributing to the characteristic aroma of malted barley. *Journal of Agricultural and Food Chemistry* 42:2880–2884.

Belitz, H. D., W. Grosch, and P. Schieberle. 2009. *Food Chemistry*, 4th ed. Berlin: Springer.

Bloem, A, A Lonvaud-Funel, and G. de Revel. 2008. Hydrolysis of glycosidically bound flavor compounds from oak wood by oenococcus oeni. *Food Microbiology* 25:99–104.

Boothroyd, E. 2013. *Investigation of the Congeners Responsible for Nutty/Cereal Aroma Character in New Make Malt Whisky*. PhD Thesis. Nottingham: University of Nottingham.

Brebu, M., and C. Vasile. 2010. Thermal degradation of lignin—a review. *Cellulose Chemistry and Technology* 44:353–363.

Campbell, I. 2003. Yeast and fermentation. In *Whisky: Technology, Production and Marketing*, edited by I. Russell, 117–150. London: Elsevier.

Coghe, S., E. Martens, H. D'Hollander, P. J. Dirinck, and F. R. Delvaux. 2012. Sensory and instrumental flavor analysis of wort brewed with dark specialty malts. *Journal of the Institute of Brewing* 110:94–103.

Conner, J. M. 2014. Maturation. In *Whisky: Technology, Production and Marketing*, edited by Inge Russell and Graham Stewart, 199–220. Oxford: Academic Press.

Conner, J. M., F. Jack, and D. Walker. 2010. 2-acetyl-1-pyrroline, a contributor to cereal and feinty aromas in scotch whisky. In *Worldwide Distilled Spirits Conference: New Horizons: Energy, Environment and Enlightenment*, edited by G. M. Walker and P. S. Hughes, 263–268. Nottingham: Nottingham University Press.

Cramer, A. C. J., D. S. Mattinson, J. K. Fellman, and B. K. Baik. 2005. Analysis of volatile compounds from various types of barley cultivars. *Journal of Agricultural and Food Chemistry* 53:7526–7531.

Daute, M., F. Jack, I. Baxter, B. Harrison, J. Grigor, and G. Walker. 2021. Comparison of three approaches to assess the flavor characteristics of scotch whisky spirit. *Applied Sciences* 11:1410.

Dolan, T. C. S. 1976. Some aspects of the impact of brewing science on scotch malt whisky production. *Journal of the Institute of Brewing* 82:177–181.

Dong, L., Y. Piao, X. Zhang, C. Zhao, Y. Hou, and Z. Shi. 2013. Analysis of volatile compounds from a malting process using headspace solid-phase micro-extraction and GC—MS. *Food Research International* 51:783–789.

Doyle, A., and J. C. Slaughter. 1998. Methionine and sulphate as competing and complementary sources of sulphur for yeast during fermentation. *Journal of the Institute of Brewing* 104:147–155.

Ferreira, I. F. M., and L. F. Guido. 2018. Impact of wort amino acids on beer flavor: A review. *Fermentation* 4:23.

Ferreira, V., R. Lopez, and J. F. Cacho. 2000. Quantitative determination of the odorants of young red wines from different grape varieties. *Journal of the Science of Food and Agriculture* 80:1659–1667.

Freeman, J., T. A. Bringhurst, A. Broadhead, J. M. Brosnan, and J. W. Walker. 1999. Amino acid turnover in distilleries. In *Proceedings of the Fifth Aviemore Conference on Malting Brewing and Distilling*, edited by I. Campbell, 267–270. London: Institute of Brewing.

Fujii, T., M. Kurokawa, and M. Saita. 1992. Studies of volatile compounds in whisky during ageing. In *Élaboration et Connaissance des Spiritueux: Recherche de la Qualité, Tradition et Innovation*, edited by R. Cantagrel, 543–547. Paris: Lavoisier.

Fujiwara, D., H. Yoshimoto, H. Sone, S. Harashima, and Y. Tamai. 1998. Transcriptional co-regulation of Saccharomyces cerevisiae alcohol acetyltransferase gene, ATF1 and delta-9 fatty acid desaturase gene, OLE1 by unsaturated fatty acids. *Yeast* 14:711–721.

Furasawa, T. 1996. *The Formation and Reactions of Sulphur Compounds During Distillation*. PhD Thesis. Edinburgh: Heriot Watt University.

Geddes, P. A., and H. L. Riffkin. 1989. Influence of lactic acid bacteria on aldehyde, ester and higher alcohol formation during scotch whisky fermentations. In *Distilled Beverage Flavor*, edited by J. R. Piggott and A. Paterson, 193–199. Chichester: Ellis Horwood.

Golston, A. M. 2021. The impact of barley lipids on the brewing process and final beer quality—a mini-review. *Master Brewers Association of the Americas—Technical Quarterly* 58:43–51.

The Good Scents Company. n.d. Accessed March 23, 2021. www.thegoodscentscompany.com/.

Harrison, B. 2007. *Peat Source and Its Impact on the Flavor of Scotch Whisky*. PhD Thesis. Edinburgh: Heriot Watt University.

Harrison, B., O. Fagnen, F. Jack, and J. Brosnan. 2011. The impact of copper in different parts of malt whisky pot stills on new make spirit composition and aroma. *Journal of the Institute of Brewing* 117:106–112.

Hastie, S. H. 1926. Character in pot still whisky. *Journal of the Institute of Brewing* 32:209–220.

Hastie, S. H., and W. D. Dick. 1928. Character in pot still whisky part 2. *Journal of the Institute of Brewing* 34:477–494.

Hasuo, T., and K. Toshizawa. 1986. Substance change and substance evaporation through the barrel during whisky aging. In *Proceedings of the Second Aviemore Conference on Malting, Brewing and Distilling*, 404–408. London: Institute of Brewing.

Hazelwood, L. A., S. L. Tai, V. M. Boer, J. H. de Winde, J. T. Pronk, and J. M. Daran. 2006. A new physiological role for Pdr12p inSaccharomyces cerevisiae: Export of aromatic and branched-chain organic acids produced in amino acid catabolism. *FEMS Yeast Research* 6:937–945.

Hill, A. E., and G. G. Stewart. 2019. Free amino nitrogen in brewing. *Fermentation* 5:22.

Hjelmeland, A. K., and S. E. Ebeler. 2015. Glycosidically bound volatile aroma compounds in grapes and wine: A review. *American Journal of Enology and Viticulture* 66:1–11.

Kew, W. 2016. Chemical diversity and complexity of scotch whisky as revealed by high-resolution mass spectrometry. *Journal of The American Society for Mass Spectrometry* 28:200–213.

Kłosowski, G., D. Mikulski, A. Rolbiecka, and B. Czupryński. 2017. Changes in the concentration of carbonyl compounds during the alcoholic fermentation process carried out with saccharomyces cerevisiae yeast. *Polish Journal of Microbiology* 66:327–334.

Kobayashi, N., H. Kaneda, Y. Kano, and S. Koshino. 1993. The production of linoleic and linolenic acid hydroperoxides during mashing. *Journal of Fermentation and Bioengineering* 76:371–375.

Kobayashi, N., H. Kaneda, Y. Kano, and S. Koshino. 1994. Behavior of lipid hydroperoxides during mashing. *Journal of the American Society of Brewing Chemists* 52:141–145.

Krishnamurthy, R. G., T. H. Smouse, B. D. Mookherjee, B. R. Reddy, and S. S. Chang. 1967. Identification of 2-pentyl furan in fats and oils and its relationship to the reversion flavor of soybean oil. *Journal of food Science* 32:372–374.

Kühbeck, F., W. Back, and M. Krottenthaler. 2006. Influence of lauter turbidity on wort composition, fermentation performance and beer quality—a review. *Journal of the Institute of Brewing* 112:215–221.

Kyraleou, M., D. Herb, G. O'Reilly, N. Conway, T. Bryan, and K. N. Kilcawley. 2021. The impact of terroir on the flavor of single malt whisk(e)y new make spirit. *Foods* 10:443.

Landaud, S., S. Helinck, and P. Bonnarme. 2008. Formation of volatile sulfur compounds and metabolism of methionine and other sulfur compounds in fermented food. *Applied Microbiology and Biotechnology* 77:1191–1205.

Le Floch, A., M. Jourdes, and P. L. Teissedre. 2015. Polysaccharides and lignin from oak wood used in cooperage: Composition, interest, assays: A review. *Carbohydrate Research* 417:94–102.

Lee, K. Y. M., A. Paterson, J. R. Piggott, and G. D. Richardson. 2001. Origins of flavor in whiskies and a revised flavor wheel: A review. *Journal of the Institute of Brewing* 107:287–313.

Lentz, M. 2018. The impact of simple phenolic compounds on beer aroma and flavor. *Fermentation* 4:20.

Lukić, I., S. Tomas, B. Miličević, S. Radeka, and D. Peršurić. 2011. Behaviour of volatile compounds during traditional alembic distillation of fermented muscat blanc and Muškat Ruža Porečki Grape Marcs. *Journal of the Institute of Brewing* 117:440–450.

Maga, J. A. 1978. Cereal volatiles: A review. *Journal of Agricultural and Food Chemistry* 26:175–178.

Masson, E., R. Baumes, C. Le Guerneve, and J. Puech. 2000. Identification of a precursor of beta-methyl-gamma-octalactone in the wood of sessile oak (Quercus petraea (Matt.) Liebl.). *Journal of Agricultural and Food Chemistry* 48:4306–4309.

Master of Malt. n.d. *Whisky or Whiskey*. Accessed March 21, 2021. www.masterofmalt. com/c/guides/whisky-or-whiskey/.

Masuda, M., and K. Nishimura. 1980. Occurrences and formation of damascenone, trans-2, 6, 6-trimethyl-1-crotonyll- cyclohexa-1, 3-diene, in alcohol beverages. *Journal of Food Science* 45:396–397.

Moir, M. 1992. The 1990 Laurence Bishop Silve medal lecture: The desideratum for flavor control. *Journal of the Institute of Brewing* 98:215–220.

Muller, S., and A. McEwan. 1998. Observations on the changes in copper concentration during the maturation of malt whisky. In *Proceedings of the Fifth Aviemore Conference on Malting, Brewing and Distilling*, 318–321. London: Institute of Brewing.

Neri, L. 2006. *The Involvement of Wild Yeast in Malt Whisky Fermentations*. PhD Thesis. Edinburgh: Heriot Watt University.

Nishimura, K., O. Masami, M. Masuda, K. Koga, and R. Matsuyama. 1983. Reactions of wood components. In *Flavor of Distilled Beverages: Origin and Development*, edited by J. R. Piggott, 241–255. Chichester: Ellis Horwood.

Noguchi, Y. 2016. *Influence of Wood Species of Casks on Matured Whisky Aroma: Identification of Unique CharacterImparted to Whisky by Casks Constructed of Japanese Oak*. PhD Thesis. Edinburgh: Heriot Watt University.

Peppard, T. L., and S. A. Halsey. 1981. Malt flavor—transformation of carbonyl compounds by yeast during fermentation. *Journal of the Institute of Brewing* 87:386–390.

Perpète, P., and S. Collin. 1999. Fate of the worty flavors in a cold contact fermentation. *Food Chemistry* 66:359–363.

Perry, D. R. 1983. Odor intensity of whisky compounds. In *Distilled beverage flavor: Recent developments*, edited by J. R. Piggot and A. Paterson, 200–207. Chichester: Ellis Horwood.

Perry, D. R. 1986. Whisky maturation mechanisms. In *Proceedings from the Second Aviemore Conference on Malting, Brewing and Distilling*, edited by I. Campbell and F. G. Priest, 409–412. London: Institute of Brewing.

Perry, D. R., A. Ford, and G. Burke. 1990. Cask rejuvenation. In *Proceedings of the Third Aviemore Conference on Malting, Brewing and Distilling*, 464–467. London: Institute of Brewing.

Philp, J. M. 1986. Scotch whisky development during maturation. In *Proceedings of the Second Aviemore Conference on Malting, Brewing and Distilling*, 148–163. London: Institute of Brewing.

Piggott, J. R., R. Sharp, and R. E. B. Duncan. 1989. *The Science and Technology of Whiskies*. Harlow: Longman Scientific and Technical.

Piornos, J. A., D. P. Balagiannis, L. Methven, E. Koussissi, E. Brouwer, and J. K. Parker. 2020. Elucidating the odor-active aroma compounds in alcohol-free beer and their contribution to the worty flavor. *Journal of Agricultural and Food Chemistry* 68:10088–10096.

Pires, E. J., J. A. Teixeira, T. Brányik, and A. A. Vicente. 2014. Yeast: The soul of beer's aroma—a review of flavor-active esters and higher alcohols produced by the brewing yeast. *Applied Microbiology and Biotechnology* 98:1937–1949.

Poisson, L., and P. Schieberle. 2008. Characterization of the key aroma compounds in an american bourbon whisky by quantitative measurements, aroma recombination, and omission studies. *Journal of Agricultural and Food Chemistry* 56:5820–5826.

Quain, D. 1989. Fermentation and its effect on flavor and aroma. *Brewers Guardian* 118:24–30.

Ramsay, C. M., and D. R. Berry. 1983. Factors influencing the aroma. *Brewing and Distilling International* (December):34–37.

Reazin, G. 1981. Chemical mechanisms of whiskey maturation. *American Journal of Enology and Viticulture* 32:283–289.

Roullier-Gall, C., J. Signoret, D. Hemmler *et al.* 2018. Usage of FT-ICR-MS metabolomics for characterizing the chemical signatures of barrel-aged whisky. *Frontiers in Chemistry* 6:29.

Rowell, R. M., R. Pettersen, J. S. Han, J. S. Rowell, and M. A. Tshabalala. 2012. Cell wall chemistry. In *Handbook of Wood Chemistry and Wood Composites*, edited by Roger M. Rowell, 35–74. Boca Raton: CRC Press. www.routledgehandbooks.com/doi/10.1201/b12487-5.

Russell, I., and G. Stewart. 2014. *Whisky: Technology, Production and Marketing*. Oxford: Academic Press.

Saerens, S. M. G., F. R. Delvaux, K. J. Verstrepen, and J. M. Thevelein. 2010. Production and biological function of volatile esters in saccharomyces cerevisiae during Fermentation. *Microbial Biotechnology* 3:165–177.

Saerens, S. M. G., F. Delvaux, K. J. Verstrepen, P. Van Dijck, J. M. Thevelein, and F. R. Delvaux. 2018. Parameters affecting ethyl ester production by saccharomyces cerevisiae during fermentation. *Applied and Environmental Microbiology* 74:454–461.

Saerens, S. M. G., K. J. Verstrepen, S. D. M. Van Laere *et al.* 2006. The saccharomyces cerevisiae EHT1 and EEB1 genes encode novel enzymes with medium-chain fatty acid ethyl ester synthesis and hydrolysis capacity. *Journal of Biological Chemistry* 281:4446–4456.

Salo, P., L. Nykanen, and H. Suomalainen. 1972. Odor thresholds and relative intensities of volatile aroma components in an artificial beverage imitating whisky. *Journal of Food Science* 37:394–398.

Schidrowitz, P. 1902. Chemistry of whisky—part 1. *Journal of the Society of Chemical Industry:* 814–819.

Schidrowitz, P., and F. Kaye. 1905. The chemistry of whisky—part 2. *Journal of the Society of Chemical Industry:* 585–589.

Schieberle, P. 1989. Formation of 2-acetyl-l-pyrroline and other important flavor compounds in wheat bread crust. In *Thermal Generation of Aromas*, edited by T. H. Parliment, R. J. McGorrin, and C. T. Ho, 268–275. Washington D.C.: American Chemical Society.

Scholtes, C., S. Nizet, and S. Collin. 2014. Guaiacol and 4-methylphenol as specific markers of torrefied malts: Fate of volatile phenols in special beers through aging. *Journal of Agricultural and Food Chemistry* 62:9522–9528.

Schwarz, P., P. Stanley, and S. Solberg. 2002. Activity of lipase during mashing. *Journal of the American Society of Brewing Chemists* 60:107–109.

The Scotch Whisky Association. n.d. *Legal Protection in the UK*. Accessed March 21, 2021. www.scotch-whisky.org.uk/insights/protecting-scotch-whisky/legal-protection-in-the-uk/.

Shortreed, W., P. Rickards, J. S. Swan, and S. Burtles. 1979. The flavor terminology of scotch whisky. *Brewers' Guardian* (November):2–6.

Slaghenaufi, D., C. Franc, N. Mora, S. Marchand, M. C. Perello, and G. de Revel. 2016. Quantification of three galloylglucoside flavor precursors by liquid chromatography tandem mass spectrometry in brandies aged in oak wood barrels. *Journal of Chromatography A* 1442:26–32.

Smit, B. A., W. J. M. Engels, and G. Smit. 2009. Branched chain aldehydes: Production and breakdown pathways and relevance for flavor in foods. *Applied Microbiology and Biotechnology* 81:987–999.

Stewart, G. G. 2017. The production of secondary metabolites with flavor potential during brewing and distilling wort fermentation. *Fermentation* 3:63.

Stewart, G. G., T. Yonesawa, and S. A. Martin. 2007. Influence of mashing conditions on fermentation characteristics of all-malt wort used to produce beer or whisky. *Master Brewers Association of the Americas Technical Quarterly* 44:256–263.

Su, C. 2003. *Fatty Acid Composition of Oils, Their Oxidative, Flavor and Heat Stabilities and the Resultant Quality in Foods.* PhD Thesis. Ames: Iowa State University.

Suomalainen, H. 1981. Yeast esterases and aroma esters in alcoholic beverages. *Journal of the Institute of Brewing* 87:296–300.

Taylor, G. T., P. A. Thurston, and B. H. Kirsop. 1979. The influence of lipids derived from malt spent grains on yeast metabolism and fermentation. *Journal of the Institute of Brewing* 85:219–227.

Thulasidas, S. 2007. *Effect of Copper on Individual Sulphur Compounds During Distillation.* MSc Thesis. Edinburgh: Heriot Watt University.

Tressl, R., D. Bahri, and B. Helak. 1983. Flavors of malts and other cereals. In *Flavor of Distilled Beverages: Origin and Development*, edited by J. R. Piggott, 9–32. Chichester: Ellis Horwood.

UK Parliament. 2009. *The Scotch Whisky Regulations 2009.* Accessed March 25, 2021. www.legislation.gov.uk/uksi/2009/2890/contents.

Vanderhaegen, B., H. Neven, H. Verachtert, and G. Derdelinckx. 2006. The chemistry of beer aging—a critical review. *Food Chemistry* 95:357–381.

Van Laere, S. D. M., K. J. Verstrepen, J. M. Thevelein, P. Vandijck, and F. R. Delvaux. 2008. Formation of higher alcohols and their acetate esters. *Cerevisia* 33:65–81.

Verstrepen, K. J., S. D. M. Van Laere, B. M. P. Vanderhaegen *et al.* 2003. Expression levels of the yeast alcohol acetyltransferase genes ATF1, Lg-ATF1, and ATF2 control the formation of a broad range of volatile esters. *Applied and Environmental Microbiology* 69:5228–5237.

Vuralhan, Z., M. A. Luttik, S. L. Tai *et al.* 2005. Physiological characterization of the ARO10-dependent, broad-substrate-specificity 2-oxo acid decarboxylase activity of saccharomyces cerevisiae. *Applied and Environmental Microbiology* 71:3276–3284.

Vuralhan, Z., M. A. Morais, S. L. Tai, M. D. Piper, and J. T. Pronk. 2003. Identification and characterization of phenylpyruvate decarboxylase genes in saccharomyces cerevisiae. *Applied and Environmental Microbiology* 69:4534–4541.

Walker, G. M., and A. E. Hill. 2016. Saccharomyces cerevisiae in the production of whisk(e)y. *Beverages* 2:38.

Wanikawa, A., K. Hosoi, and T. Kato. 2000. Conversion of unsaturated fatty acids to precursors of γ-lactones by lactic acid bacteria during the production of malt whisky. *Journal of the American Society of Brewing Chemists* 58:51–56.

Wanikawa, A., K. Hosoi, T. Kato, and K. Nakagawa. 2002. Identification of green note compounds in malt whisky using multidimensional gas chromatography. *Flavor and Fragrance Journal* 17:207–211.

Wanikawa, A., K. Hosoi, H. Shoji, and K. Nakagawa. 2001. Estimation of the distribution of enantiomers of γ-decalactone and γ-dodecalactone in malt whisky. *Journal of the Institute of Brewing* 107:253–259.

Wanikawa, A., K. Hosoi, I. Takise, and T. Kato. 2012. Detection of γ-lactones in malt whisky. *Journal of the Institute of Brewing* 106:39–44.

Wanikawa, A., H. Shoji, K. Hosoi, and K. Nakagawa. 2002. Stereospecificity of 10-hydroxystearic acid and formation of 10-ketostearic acid by lactic acid bacteria. *Journal of the American Society of Brewing Chemists* 60:14–20.

Wasowicz, E., A. Gramza, M. Hêœ *et al.* 2004. Oxidation of lipids in food. *Polish Journal of Food and Nutrition Sciences* 13:87–100.

Waterhouse, A. L., and J. P. Towey. 1994. Oak lactone isomer ratio distinguishes between wine fermented in American and French oak barrels. *Journal of Agricultural and Food Chemistry* 42:1971–1974.

Watson, D. C. 1983a. A laboratory apparatus for distillation research. In *Current Developments in Malting, Brewing and Distilling*, edited by F. G. Priest and I. Campbell, 249. London: Institute of Brewing.

Watson, D. C. 1983b. Factors influencing the congener composition of malt whisky new make spirit. In *Flavor of Distilled Beverages: Origin and Development*, edited by J. R. Piggott, 79–92. Chichester: Ellis Horwood.

Watts, S. 2005. *Thiamine: A Potential Precursor of Flavor Active Compounds in Scotch Malt Whisky*. PhD Thesis. Edinburgh: Heriot Watt University.

Wei, W., D. D. Handoko, L. Pather, L. Methven, and J. S. Elmore. 2017. Evaluation of 2-acetyl-1-pyrroline in foods, with an emphasis on rice flavor. *Food Chemistry* 232:531–544.

Wilson, C. A. 2008. *The Role of Water Composition on Malt Spirit Quality*. PhD Thesis. Edinburgh: Heriot Watt University.

Wilson, N. R. 2008. *The Effects of Lactic Acid Bacteria on Congener Composition and Sensory Characteristics of Scotch Malt Whisky*. PhD Thesis. Edinburgh: Heriot Watt University.

Yu, J., S. Huang, J. Dong *et al.* 2014. The influence of LOX-less barley malt on the flavor stability of wort and beer. *Journal of the Institute of Brewing* 120:93–98.

Section II

Fermented Beverages

Chapter 5

Volatile Compounds Formation in Sparkling Wine

Juliane Elisa Welke, Bruna Dachery, Lucas Dal Magro,
Karolina Cardoso Hernandes, and Cláudia Alcaraz Zini

CONTENTS

5.1 INTRODUCTION

Paramount importance is given to volatile compounds of sparkling wines, as they are associated with their organoleptic characteristics. Volatile compounds contribute to flavor, which is defined as the sensation arising from the integration or interplay of signals produced as a consequence of tactility (mouthfeel), taste (gustatory), and smell (olfactory) (International Organization for Standardization 2008; Welke *et al.* 2021). Odor and aroma are olfactory perceptions from volatile

DOI: 10.1201/9781003129462-7

components that enter the nasal passage through the nose (orthogonally) and at the back of the throat (retronasally), respectively. The aroma perception requires tasting (International Organization for Standardization 2008). Overall, a volatile compound may be odor-active and contributes to the aroma when found at a concentration above the odor perception threshold (defined as the lowest concentration capable of producing an olfactory sensation and that can be detected by 50% of the assessors) (ASTM International 2012; International Organization for Standardization 2002). However, the interaction of multiple components present in subthreshold concentrations has been proven to contribute indirectly to fruity expression (McKay *et al.* 2020).

The volatile profile results from the combination of factors such as grape characteristics (variety and degree of ripeness), fermentation conditions (spontaneous, use of commercial microorganisms, yeast strain, time, temperature, and type of container), methods of preparation (Traditional, Charmat, Asti, Transfer, or Ancestral), maturation and aging (Carlin *et al.* 2016; Muñoz-Redondo *et al.* 2020; López de Lerma *et al.* 2018). It is worth mentioning that the volatile composition is also strongly influenced by the *terroir*, defined as an area in which collective knowledge of the interactions between the physical and biological environment and applied vitivinicultural practices, provides distinctive characteristics for the products (OIV 2020).

The origin of volatile compounds and their sensory importance is the object of study mainly in still wines (Lin, Massonnet, and Cantu 2019; Ilc, Werck-Reichhart, and Navrot 2016; Styger, Prior, and Bauer 2011). The literature on the relevance of these attributes for sparkling wines is incipient, considering the wide range of types of sparkling wines produced around the world. In this chapter, an overview of the production methods is given, highlighting the particular characteristics of each type of sparkling wine and the factors that govern the formation of volatile compounds.

5.2 OVERVIEW OF THE METHODS USED IN THE PRODUCTION OF SPARKLING WINES

The knowledge of the steps involved in the preparation of each type of sparkling wine is important in order to understand the origin and the role of volatile compounds in the consumer's perception. The main technique used to produce sparkling wine involves a second alcoholic fermentation induced by adding sugar and yeast to a base wine, as occurs in Traditional (also known as Classical or Champenoise), Charmat, and Transfer methods. Another way to produce sparkling wine is through a single fermentation, as performed by the Ancestral and Asti methods. Figure 5.1 shows the steps of sparkling wines production by the Traditional, Charmat, Transfer, Ancestral, and Asti methods. All methods start by obtaining a must from grape pressing and are followed

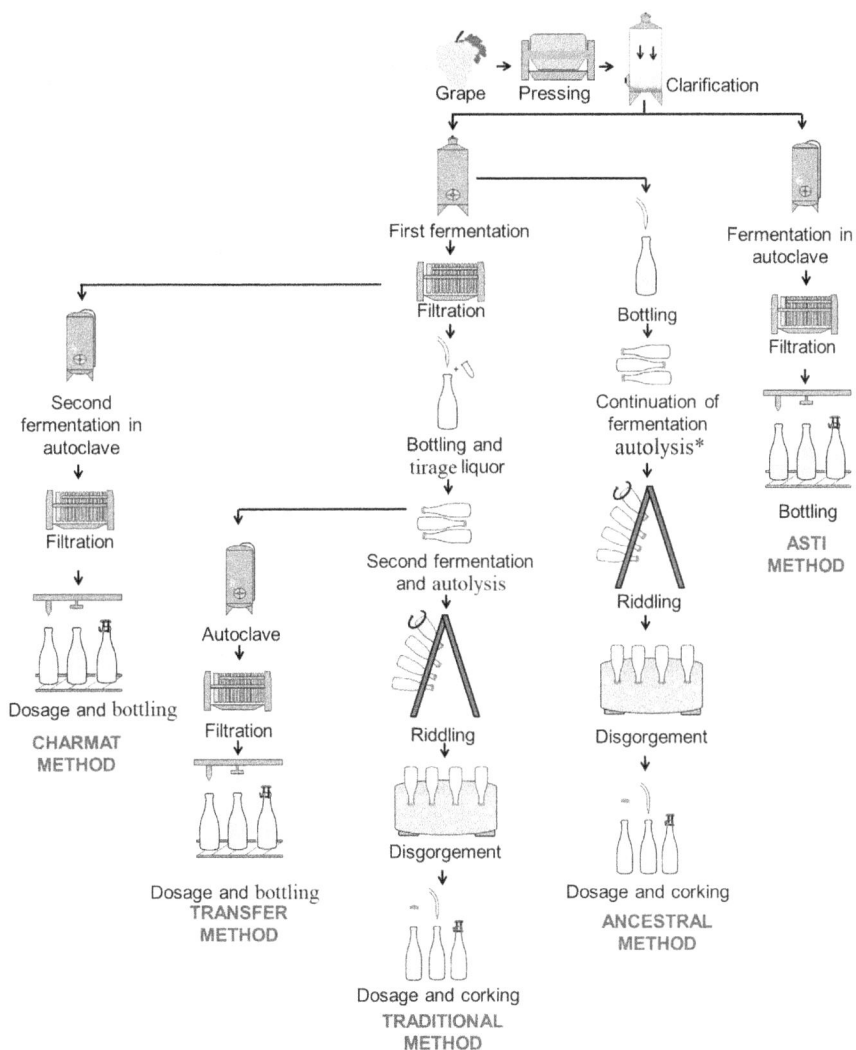

Figure 5.1 Stages of Sparkling Wine Production by the Traditional, Charmat, Transfer, Ancestral, and Asti Methods.

Note: *Autolysis can occur in the Ancestral method according to the criteria established by each region.

by clarification. This step is carried out with fining agents (e.g., bentonite) that remove substances from the must in order to enhance clarity, taste, and/or stability (Ribéreau-Gayon *et al.* 2006). A more detailed explanation about the methods of production of sparkling wines can be found in several publications (Jackson 2020; Jeandet *et al.* 1997; Buxaderas and López-Tamames 2012).

Traditional and Charmat are the most used methods worldwide. Therefore, studies have been focused mainly on the influence of these methods on the quality of sparkling wines (Jeandet *et al.* 1997; Ubeda *et al.* 2016; Caliari *et al.* 2015).

For the production of the base wine used in methods that include two fermentations (Transfer, Traditional, and Charmat), the grape must is intended for alcoholic fermentation in tanks with temperature control (15–18°C). Fermentation is completed when the wine reaches around 10.5% of alcohol. The base wine is filtered and tirage liquor (yeast and sugar) is added for the second fermentation in autoclaves (Charmat method) or in bottles (Traditional and Transfer methods) (Jackson 2020).

The Traditional method was developed in the Champagne region (France) by the monk Dom Pérignon in the 17th century (Jeandet *et al.* 1997). The second fermentation takes place in the bottle, where carbon dioxide is retained, after the addition of yeast and sugar. A maturation stage usually occurs, starting between three and four months after the end of the second fermentation (Alexandre and Guilloux-Benatier 2006). Maturation refers mainly to the autolysis process, which is defined as the degradation of the yeast cell walls (lees). Sparkling wines like Champagne (France), Cava (Spain), Franciacorta (Italy), Winzersekt (Germany), Crémant (France), Cap Classique (South Africa), and Trento (Italy) are produced by the Traditional method, with different types of grapes, and under criteria established by the designations of origin.

The Charmat method emerged in the 20th century, when the French Eugène Charmat patented a method developed by the Italian Federico Martinotti, which differs from the Traditional method by conducting the second fermentation in large pressurized containers, called sealed tanks or autoclaves. This simpler and less laborious method was developed aiming to produce a lower-cost, sparkling wine if compared to the one of the Traditional method (Buxaderas and López-Tamames 2012). The fermentation in a sealed tank preserves varietal aromas since the sparkling wine is filtered (the autolysis stage does not occur) after the end of the second fermentation (Torchio *et al.* 2012). This characteristic is advantageous when aromatic grape varieties such as Muscat and Riesling, among others, are intended for winemaking. Currently Prosecco (Denominazione di Origine Controllata [DOC] produced in Friuli Venezia Giulia and Veneto regions or Denominazione di Origine Controllata e Garantita [DOCG] produced in Conegliano Valdobbiadene, an area of the Veneto region) is the main sparkling wine produced by the Charmat method, representing 66% of Italian production (OIV 2020).

In the Transfer method, once the second fermentation and autolysis are completed inside the bottle, the sparkling wine is transferred to a pressurized tank, filtered, and bottled. This method is a hybrid of the Traditional and tank methods, allowing the sparkling wine to gain the benefits of the aging on lees without the expense or time of riddling and disgorgement. Although laborious, it is occasionally used to elaborate sparkling wines sold in bottles of different sizes (375 mL, 1.5 L, or 6 L) than the standard one (750 mL) (Ribéreau-Gayon *et al.* 2006).

Contrary to the Traditional, Charmat, and Transfer methods, Ancestral (also called Rural) and Asti methods produce sparkling wines by a single fermentation. Only some French winegrowing regions officially use the Ancestral method, including Gaillac (Southwestern France), Clairette de Die (Rhône Valley), and Bugey et Cerdon (Bugey). Even though each region presents unique peculiarities, the objective consists of bottling the base wine without having a complete fermentation, so that it continues inside the bottle (without adding yeast and sugars). When fermentation in the bottle finishes, autolysis can occur (or not) according to the criteria established by each region (Jeandet *et al.* 1997).

A first report on a sweet aromatic wine is attributed to Giovan Battista Croce in the 17th century (Croce 2004), but the first sparkling Asti is credited to Carlo Gancia, around 1870, with the aim to inhibit excess refermentation in the bottle (Montella 2012). For the Asti method, the grape must is placed in autoclaves to carry out alcoholic fermentation. When the wine reaches 6% of alcohol content, the tank valve is closed so that carbon dioxide is retained. Once 5 bar pressure is reached, fermentation is interrupted by the cooling of the sparkling wine to approximately 0°C. Therefore, only a part of the sugars present in the grape is fermented, and varietal aromas are preserved. This method is used to produce Asti Spumante DOCG (Italy) and Moscatel (Brazil) (Ribéreau-Gayon *et al.* 2006).

5.3 THE IMPORTANCE OF VOLATILE COMPOUNDS FOR THE QUALITY OF SPARKLING WINES

The use of analytical techniques, including the use of comprehensive two-dimensional gas chromatography with mass spectrometry detection (GC×GC/MS) (Carlin *et al.* 2016; Soares *et al.* 2015; Welke *et al.* 2014), monodimensional gas chromatography with mass spectrometry detection (GC/MS) (Martínez-García *et al.* 2021; Korenika *et al.* 2020; Muñoz-Redondo *et al.* 2020), gas chromatography with olfactometric detection (Torrens *et al.* 2010; Ubeda *et al.* 2019), and electronic nose (Martínez-García *et al.* 2021) associated with sensory analysis (Ubeda *et al.* 2019; López de Lerma *et al.* 2018), allows an extensive characterization of sparkling wines produced by different methods, grape varieties, yeast strains, and times of maturation, contributing to a better understanding the role of volatile compounds. Table 5.1 summarizes the volatile compounds identified as markers or as the most discriminative compounds for different types of sparkling wines.

TABLE 5.1   LITERATURE DATA REGARDING THE CHARACTERIZATION OF THE VOLATILE PROFILE OF COMMERCIAL SPARKLING WINES.

Grapes / Traditional method	Country	Markers or the Most Discriminative Compounds	Techniques[a]	References
Cava (50% Macabeo, 30% Xarel-lo and, 20% Parellada)	Spain	**Aging for 15 months:** Acetate ester: hexyl acetate, 2-phenylethyl acetate, and isoamyl acetate Ethyl ester: ethyl hexadecanoate Lactone: caprolactone **Aging for 24 months:** Aldehyde: benzaldehyde Acetate ester: ethyl-2-phenylacetate Ethyl ester: ethyl isobutanoate Lactone: decalactone	GC-FID GC/MS electronic-nose	(Martínez-García et al. 2021)
Chardonnay, Pinot noir, Portugizer, Kraljevina, and Manzoni Bianco grape	Croatia	**Aging for 9 months:** C13-norisoprenoid: 1,1,6-trimethyl-1,2-dihydronaphthalene (TDN) Diethyl esters: diethyl succinate and diethyl glutarate	GC/MS	(Korenika et al. 2020)
Cava, Champagne, and sparkling wine produced in Andalusia (without protected designations of origin [PDO])	Spain and France	**Aging for 9–15 months:** C13-norisoprenoid: TDN Monoterpenes: ß-citronellol and α-terpineol, β-cyclocitral, geranyl acetate, cis-citral, camphor, γ-terpinene, and terpinen-4-ol Sesquiterpene: trans-nerolidol **Aging for 15–24 months:** Sesquiterpenes: cis-nerolidol and trans-nerolidol **Aging for over 24 months:** C13-norisoprenoid: β-damascenone Monoterpenes: trans-linalyl oxide, eucalyptol, citronellal and terpinen-4-ol Sesquiterpenes: neryl acetate	GC/MS	(Muñoz-Redondo et al. 2020)

País	Chile	**Aging for 12 months:** Alcohols: hexanol and 3-hexenol C13-norisoprenoid: vitispirane Esters: diethyl succinate, ethyl lactate, and ethyl isovalerate Furan: furfural	GC/MS GC-O Descriptive sensory analysis	(Ubeda et al. 2019)
Cava (30% Macabeo, 40% Parellada, and 30% Xarel-lo)	Spain	**Aging for 32 months:** Acids: octanoicacid and decanoic acid Alcohol: hexanol (C6 compounds) C13 norisoprenoids: vitispirane and TDN	GC/MS Descriptive sensory analysis	(López de Lerma et al. 2018)
White sparkling	Italy	**Impact of Saccharomyces cerevisiae strains:** Sparkling wines obtained F10471 and F10477 strains were characterized by: Esters: 3-methylbut-1-yl ethanoate, ethyl ethanoate, ethyl octanoate, ethyl decanoate and ethyl dodecanoate	GC/MS	(Di Gianvito et al. 2018)
Cava (60% Macabeo and 40% Chardonnay)	Spain	**Changes in aroma compounds during fermentation carried out under nonpressure condition and CO_2 overpressure condition** Esters: ethyl isobutanoate, ethyl octanoate, ethyl dodecanoate, ethyl tetradecanoate, ethyl butanoate, ethyl isobutanoate and hexyl acetate Alcohols: 1-hexanol Phenol:4-ethenyl-2-methoxyphenol Furan: 2(5H)-furanone	GC-FID GC/MS	(Martínez-García et al. 2017)

(Continued)

TABLE 5.1 CONTINUATION

Grapes	Country	Markers or the Most Discriminative Compounds	Techniques[a]	References
Pedro Ximenez grapes	Spain	**Aging for 9 months:** Ethyl esters: isovalerate, isobutyrate, and 2-methylbutyrate	GC/MS	(Ruiz-Moreno et al. 2017)
Trentodoc wine (Certified Brand of Origin, Italy) Franciacorta (Controlled and Guaranteed Denomination of Origin, DOCG, Italy)	Italy	**Average time of aging: 36 and 48 months for Trentodoc and Franciacorta, respectively.** Trentodoc: C6 compounds: hexanol, cis- and trans-3-hexenol trans-2-hexenol and hexyl acetate Ethyl ethers monoterpenes: linalyl, neryl geranyl and α-terpenyl ethyl ether Monoterpenes: cis-rose oxide, linalool, nerol oxide, α-terpineol, 2,6 dimethyl 1,7 octadien 3-oland a p-menth-1-en-9-al **Franciacorta:** Diethyl esters: glutarate, malate and malonate Sesquiterpene: γ-eudesmol Sulfur compounds: 3-ethoxy thiophene, ethyl 2-methylthioacetate, thieno[2,3-b] thiophene, 2-methylthiophene,2-thiophenecarboxaldehyde and tetrahydrothiophen-3-one	GC×GC/TOFMS	(Carlin et al. 2016)
Chardonnay	Brazil	**Aging for 9 months:** Alcohols: 4-butoxy butanoland1-propanol Aldehydes: 3-phenyl-2-propenal, nonanal, methional, and undecanal Acids: acetic, 2-ethylhexanoicand butanoic C13-norisoprenoids: TDN, vitispirane and β-damascenone Ethyl esters: laurate, 2-hydroxybutanoate, decanoate, 2-hydroxypropanoate, and pentanoate Ketones: acetoin and diacetyl Phenols: 4-vinylguaiacol and 4-ethylbhenol	GC×GC/MS	(Welke et al. 2014)

Classical varieties: Pinot Noir, Pinot Gris, Chardonnay, Sauvignon Blanc, and Riesling Renano. Innovative varieties: Moscato Embrapa, Niagara, Villenave, Goethe, and Manzoni Bianco	Brazil	**Aging for 18 months:** Markers were identified in the sparkling wines elaborated with the following grape varieties: Moscato Embrapa: isoamyl acetate (acetate ester) and geraniol (monoterpene) Villenave: ethyl octanoate (ethyl ester) Riesling Renano: terpene: linalool (monoterpene) Niagara: linalol oxide (monoterpene)	GC/MS GC-FID	(Caliari et al. 2014)
Brachetto	Italy	**Aging for 12 months:** Two types of sparkling wines were elaborated according to the Brachetto d'Acqui DOCG Disciplinary of Production: a sweet lightly sparkling wine (final bottle pressure 1.7 bar, called Tapporaso) and a sweet fully sparkling wine (final bottle pressure > 3.0 bar, called Spumante) Lightly: geraniol, cis-pyran linalool oxide and cis-furan linalool oxide (monoterpenes) Fully: trans-pyran linalool oxide, nerol, citronellol and 2,6-dimethyl-3,7-octadien-2,6-diol (monoterpenes)	GC/MS Sensory preference (olfactory judgment)	(Torchio et al. 2012)

(Continued)

TABLE 5.1 CONTINUATION

Grapes	Country	Markers or the Most Discriminative Compounds	Techniques[a]	References
Cava—Macabeu, Xarel-lo, and Parellada	Spain	**Aging for 14 or 24 months:** Acetate esters (phenyl ethyl acetate, octyl acetate and hexyl acetate) (concentration decreased during aging) Alcohols: benzylic alcohol and 2-phenylethanol C13-norisoprenoids: vitispiranes and TDN Ethyl esters: ethyl lactate and diethyl succinate Terpene: linalool Furans: furfural, 5-methylfurfural and acetyl furan Sulfur compounds: 2-methyl-3-furanthiol and furfuryl thiol	GC/MS GC-FID GC-O	(Torrens et al. 2010)
Rosé sparkling Cava—Trepat and Monastrell White Cava—Xarel-lo: Macabeo: Parellada	Spain	**Impact of Trepat and Monastrell red grape varieties on nitrogen and volatile composition rosé sparkling Cava wines:** Compounds that allowed distinction of Cavas manufactured with the Monastrell: Esters: ethyl acetate, ethyl octanoate, ethyl decanoate and ethyl lactate Acids: octanoic acid Aldehyde: acetaldehyde Alcohol: methanol **Aging on lees for 9, 12, 15, and 18 months:** Compounds affect by aging: Ester: ethyl acetate Alcohols: 1-propanol, isobutanol, 3-methyl-1-butanol	GC-FID Descriptive sensory analysis	(Pozo-Bayón et al. 2010)

Cava-Macabeu, Xarel-lo and Parellada (1:1:1) Rosécava (Red Trepat variety)	Spain	**Aging for 27 months:** Acetate esters: 2-phenylethyl and isoamyl acetate (concentration decreased during aging) C6 compound: Hexyl acetate C13 norisoprenoids: vitispiranes and TDN Ethyl ester: diethyl succinate	GC/MS	(Riu-Aumatell et al. 2006)
Champagne (Chardonnay 50% and Pinot Noir 50%)	France	**Aging ranged from 0 to 27 years:** Sulfur compounds: 2-furanmethanethiol, benzene methane thiol and ethyl3-mercaptopropionate Furan: furfural	GC/MS GC-O GC-FDP	(Tominaga et al. 2003)
Asti				
Moscato Bianco and Moscato R2	Brazil	**Evolution of volatile compounds during the elaboration** Terpenes: limonene, 4-terpineol, terpinolene, citronellol, α-terpineol, linalool, hotrienol, nerol, and nerol oxide Alcohols: 1-nonanol, 2-phenylethanol and 1-hexanol Esters: ethyl octanoate, ethyl decanoate and hexyl acetate	GC×GC/TOFMS GC/MS	(Soares et al. 2015)
Moscato Bianco, Moscato Giallo	Brazil	**Characterization of the volatile profile** Compounds that allowed distinction between Moscato Giallo and Moscato Bianco samples: Monoterpenes: α-terpineol, linalool, citronellol, nerol oxide, p-mentha-1,5-dien-8-ol, linalool oxide, geranyl acetone, hotrienol, Z-ocimenol, and terpinolene	GC×GC/TOFMS GC/MS	(Nicolli et al. 2015)

(Continued)

TABLE 5.1   CONTINUATION

Grapes	Country	Markers or the Most Discriminative Compounds	Techniques[a]	References
Comparison of various methods				
Pinot Noir, Chardonnay, Semillon	Chile	**Volatile profile characterization of sparkling wines produced by Traditional and Charmat methods** Compounds that allowed distinction between Traditional and Charmat methods: Esters: ethyl 2-methyl-butyrate, ethyl isovalerate, diethyl succinate, diethyl 2-hydroxy-3-methylbutanedioate, and diethyl malate Alcohols: isobutanol, 2-methyl-1-butanol and 1-undecanol Ketone: acetoin Terpenes: *cis*- and *trans*-linalool oxides and eudesmol C13-norisoprenoid: TDN Phenol: 4-vinylguaiacol and coumaran	GC/MS	(Ubeda et al. 2016)
Moscato Giallo	Brazil	**Effect of the Traditional, Charmat, and Asti method of production on the volatile composition** Traditional method: Monoterpenes: citronellol, linalool and geraniol Acids: hexanoic acid, octanoic acid and decanoic acid, Esters: isoamyl acetate and ethyl octanoate Charmat: Alcohols: hexan-1-ol and hotrienol Asti method: Monoterpenes: linalool oxide A (*trans*-furan), linalool oxide B (*cis*-furan) and linalool oxide D (*trans*-pyran)	GC-FID GC/MS	(Caliari et al. 2015)

[a]GC: gas chromatography; GCx: comprehensive two-dimensional gas chromatography; FID: flame ionization detector; MS: mass spectrometry detector; FPD: flame photometric detector; TOFMS: time-of-flight mass spectrometry detector.

According to origin, volatile compounds can be classified as:

- *Varietals*: compounds exclusively from grapes;
- *Fermentatives*: compounds from alcoholic and malolactic fermentation;
- *Postfermentatives*: compounds from the maturation and aging processes (Styger, Jacobson, and Bauer 2011; Lin, Massonnet, and Cantu 2019).

The association between varietal, fermentative, and postfermentative compounds has a direct impact on consumers' appreciation of sparkling wines (Ilc, Werck-Reichhart, and Navrot 2016). Therefore, the knowledge of the origin of these compounds and their behavior during the elaboration stages is an essential tool to guarantee the production of high-quality sparkling wines.

5.3.1 Varietal Compounds

Among the varietal compounds, monoterpenes are the main markers for sparkling wines; however, sesquiterpenes and C13-norisoprenoids may also play an important role (Alessandrini *et al.* 2017). The main metabolic pathways leading to the biosynthesis of varietal volatile compounds of sparkling wines produced by different methods are summarized in Figure 5.2.

These compounds accumulate in the pulp and mainly in the skin during grape ripening. Monoterpenes, sesquiterpenes, and C13-norisoprenoids are present as aroma precursors, as they are bound to sugar molecules. Therefore, the release of these glycosylated compounds depends on the glycosidases produced by yeasts during fermentation or autolysis. Terpenes and C13-norisoprenoids are derived from methylerythritol 4-phosphate (MEP) and mevalonic acid (MVA) pathways. In both pathways, terpenes are formed from the two universal precursors called isopentenyl diphosphate (IPP) and dimethylallyl diphosphate (DMAPP). The MEP pathway produces both IPP and DMAPP from pyruvate and glyceraldehyde 3-phosphate, and the MVA pathway synthesizes IPP from acetyl CoA, and IPP is then converted into DMAPP by IPP isomerases (Figure 5.2a) (Carrau, Boido, and Dellacassa 2008).

Varietal aroma of sparkling wines mainly comes from monoterpenes (terpenes containing two units of isoprene, 10 carbons), including linalool (Carlin *et al.* 2016), geraniol (Caliari *et al.* 2014), nerol (Torchio *et al.* 2012), and α-terpineol (Muñoz-Redondo *et al.* 2020), among others (Table 5.1), which have a high odor impact contributing to the floral, fruity or citrus characteristics (Dziadas and Jeleń 2010; Yang *et al.* 2019; Sun *et al.* 2020). As monoterpenes, sesquiterpenes (terpenes containing three isoprene units, 15 carbons) have also been reported in sparkling wines (Table 5.1), mainly nerolidol, neryl acetate (Muñoz-Redondo *et al.* 2020), and y-eudesmol (Carlin *et al.* 2016), which may contribute floral notes to aroma.

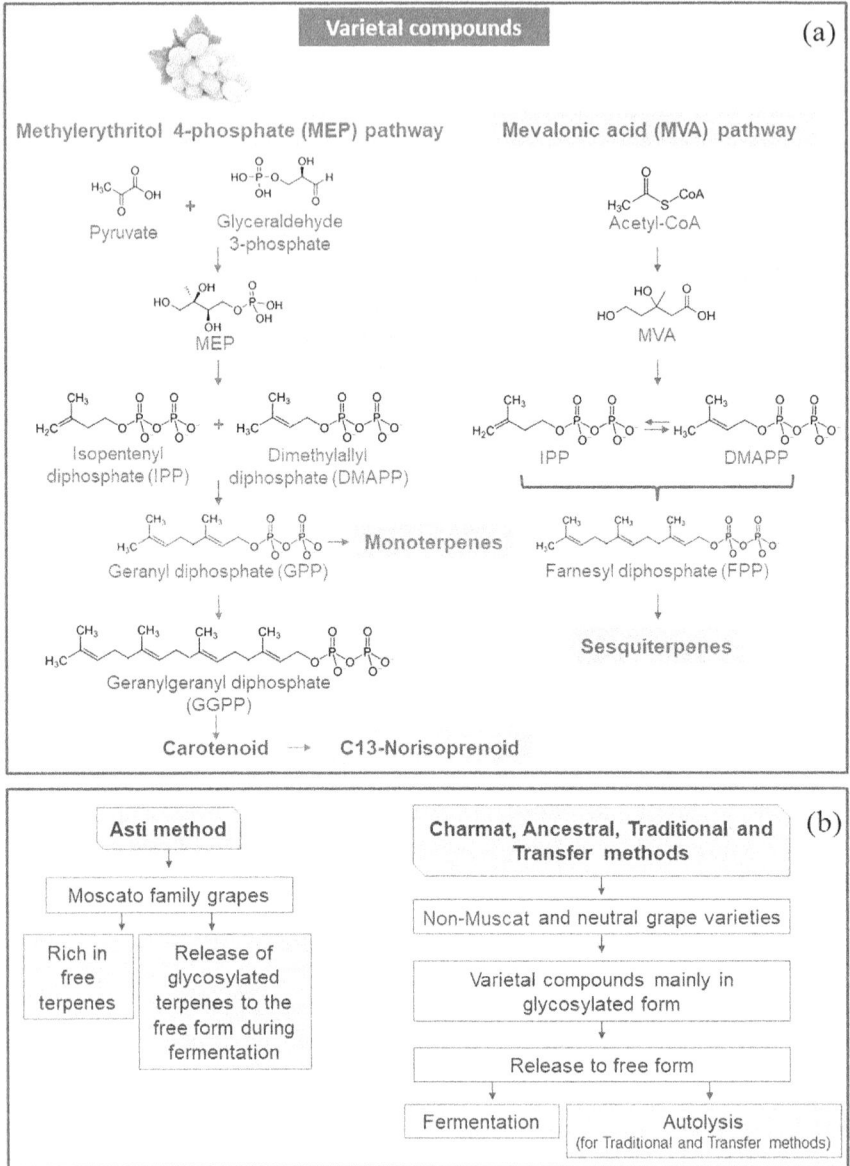

Figure 5.2 Overview of the Metabolic Pathways Leading to the Biosynthesis of (a) Varietal Compounds and (b) the Expression of the Varietal Aroma from the Different Methods of Sparkling Wine Production.

TABLE 5.2   AROMATIC POTENTIAL OF GRAPE VARIETIES ACCORDING TO THE CONCENTRATION OF MONOTERPENES.

Aroma Potential	Monoterpene Content (mg L^{-1})	Grape Variety
Intense	4–6	**Muscat varieties:** Muscat of Alexandria, Muscat Hamburgo, Muscat de Frontignan, Muscat Blanc à PetitsGrains, Muscat Ottonel
Medium	1–4	**Non-Muscat varieties:** Gewurztraminer, Riesling, Sylvaner, Traminer, Muller-Thurgau
Neutral	< 1	**Neutral varieties:** Chardonnay, Chenin Blanc Sauvignon Blanc, Semillon, Trebbiano, Glera, Verdejo, Clairette, Pinot Noir

The concentration of terpenes varies according to the grape variety. Muscat grape varieties have an intense aroma potential due to their high monoterpene content. Table 5.2 shows the aroma potential of grape varieties used for the production of sparkling wines around the world, according to the monoterpene content (major class of varietal compounds) reported by Mateo and Jiménez (2000), Moreno-García and Peinado (2012), Jones *et al.* (2014), Yuan and Qian (2016), Alessandrini *et al.* (2017). In addition to the grapes mentioned in Table 5.2, other untraditional varieties have been introduced in the production of sparkling wines. Ribolla Gialla is a promising white grape variety that has recently been used for the production of premium sparkling wines, which have received good appreciation in international wine markets (Voce *et al.* 2019). Grillo (Alfonzo *et al.* 2020) and Maresco (Tufariello *et al.* 2019) are innovative grape varieties proposed for sparkling wine production due to their high amount of total acidity and malic acid content, as well as the low pH that improves the quality of sparkling wines.

Ethyl ethers of monoterpenes as linalyl, neryl geranyl, and α-terpinyl ethers were reported for the first time in sparkling wines by Carlin *et al.* (2016). These compounds are formed during fermentation and/or autolysis of Trentodoc wine (Certified Brand of Origin, Italy) and Franciacorta (Controlled and Guaranteed Denomination of Origin, DOCG, Italy) sparkling wines. Floral (linalyl and neryl) and fruity (geranyl) notes may be attributed to these ethyl ethers, while the contribution of α-terpinyl ethyl ether to the sensory characteristics of these sparkling wines has not yet been studied.

C13-norisoprenoids are a class of volatile compounds derived from the oxidative degradation of carotenoids, which are formed from geranyl diphosphate (GGPP). Like monoterpenes, carotenoids are synthesized via the MEP pathway. Some of these compounds have great relevance to the aroma of sparkling wines. Chardonnay (*Vitis vinífera*) is the most used grape for the production of sparkling wines in the world, especially by the Traditional method. The varietal profile of this grape presents about 70% of C13-norisoprenoids. During fermentation and in the postfermentative stage, the release of free aroma compounds from the glycoside precursors occurs, contributing to the floral and fruity notes (Moreno-García and Peinado 2012). β-damascenone is reported to be responsible for the typicality of sparkling wines produced with Chardonnay grapes using the Traditional method (Ganss *et al.* 2011; Muñoz-Redondo *et al.* 2020). Two other C13-norisoprenoids, vitispirane and 1,1,6-trimethyl-1, 2-dihydronaphthalene (TDN), have been identified as markers of sparkling wine produced by the Traditional method (Riu-Aumatell *et al.* 2006; Welke *et al.* 2014; Torrens *et al.* 2010; López de Lerma *et al.* 2018) and will be discussed in Section 5.4.

The Asti sparkling wines present the characteristic varietal aromas due to grapes of intense aroma potential (Lin, Massonnet, and Cantu 2019). For sparkling wines made by the Asti process such as Asti Spumanti (Italy) and Moscatel Espumante (Brazil) grapes from the Moscato family (*V. vinífera*) are used, which are rich in free terpenes (Soares *et al.* 2015; Sun *et al.* 2020). In addition, glycosylated forms of terpenes are released to the free form during fermentation (Figure 5.2b). Sparkling wines made with Moscato Giallo grapes by the Asti method have a higher concentration of linalool, nerol, geraniol, α-terpineol, hotrienol, and rose oxide than those produced by other methods (Caliari *et al.* 2015). These compounds impart fruity and floral characteristics to the sparkling wines, especially the linalool, which is the major terpene of the Moscato family grapes (Lin, Massonnet, and Cantu 2019; Soares *et al.* 2015).

The Charmat method has been carried out with non-Muscat or neutral grape varieties. Prosecco sparkling wine is produced following this method and has a significant sensory influence of varietal aromas (Alessandrini *et al.* 2017). Prosecco is produced with the Glera grape (*V. vinífera*), a white variety native to the Veneto region (Italy) that is grown specifically for the production of this sparkling wine (Carlin *et al.* 2016). The varietal compounds (monoterpenes and C13-norisoprenoids) of the Glera grape are found mainly in glycosylated form and in low concentrations when compared to other varieties, and for this reason it is considered a neutral variety. During fermentation, these compounds are released to the free form (Figure 5.2b). Geraniol, *cis*-8-hydroxy-linalool, linalool, and 7-hydroxy-geraniol have been found as the most representative terpenes of Glera grapes, while their main norisoprenoids are 3-oxo-α-ionol and vomifoliol. Other important varietal compounds found in Glera grapes are volatile benzenoids (present one benzene ring). Benzyl alcohol and 2-phenyl ethanol compounds are the main representatives of this group, which impart fruity and floral aroma notes, respectively. These compounds are originated specifically from phenylalanine that is metabolized via the Ehrlich pathway (Alessandrini *et al.* 2017).

For Ancestral, Traditional, and Transfer methods, non-Muscat and neutral grape varieties have also been used, in which varietal compounds are present mainly in glycosylated form. Traditional and Transfer methods include the autolysis step that may result in release of the form of free glycosylated varietal compounds, in addition to the hydrolysis of glycosidic bonds that occurs during fermentation (Figure 5.2b). This step may also be carried out in the Ancestral method, depending on regional criteria. The autolysis role has been studied mainly regarding the Traditional method and will be discussed in Section 5.4.

Table 5.1 provides a detailed review of the literature on the characterization of the volatile profile of sparkling wines. Sparkling wines produced with a limited scope of grape varieties have been studied, including mainly Macabeo, Xarel-lo and Parellada, Chardonnay, Pinot Noir, Pais, Pedro Ximenez, and Moscatel. Considering that each wine-growing region focuses its production on the grapes that best adapt to the characteristics of that specific region, scientific data on the volatile profile of sparkling wines produced with nontraditional grape varieties are needed.

5.3.2 Fermentative Compounds

Although varietal aromas are important for sparkling wines, most of their aroma compounds are derived from alcoholic fermentation due to the activity of yeasts (*Saccharomyces* and non-*Saccharomyces*) and bacteria (*Oenococcus* genera) during alcoholic and malolactic fermentation, respectively (Lin, Massonnet, and Cantu 2019).

Saccharomyces cerevisiae is typically used in sparkling wine production due to its favorable characteristics of high fermentative power (can ferment up to alcohol strength of 13–15%), alcohol tolerance, and low production of volatile acidity (Ivit *et al.* 2018). Several non-*Saccharomyces* yeasts are present on grape berries, affecting fermentation and wine composition. Considered in the past as undesirable or as spoilage microorganisms, some of these non-*Saccharomyces* yeasts are recognized by the positive effect to the aroma profile and complexity of wines (Marcon *et al.* 2018). *Torulaspora delbrueckii* is probably the non-*Saccharomyces* yeast most frequently used for wine fermentation. The sequential inoculation of *T. delbrueckii* and *S. cerevisiae* increases glycerol concentration, reduces volatile acidity, and gives a positive effect on the foaming properties of sparkling wine (González-Royo *et al.* 2015). Furthermore, the amounts of the major ethyl esters responsible for fruity aromas (ethyl hexanoate, ethyl octanoate, ethyl decanoate, and ethyl 9-decenoate), and of γ-butyrolactone (cooked peach, coconut, caramel, toasty), furfural (caramel), and β-damascenone (sweet, fruity) were more abundant in *S. cerevisiae* wines (single or mixed inoculated) than in those made with *T. delbrueckii* (Velázquez *et al.* 2019).

Malolactic fermentation has not been a common practice in the production of sparkling wines. However, in the production of Champagne (France), malolactic fermentation has been carried out since ancient times in the base wine through the inoculation of bacteria (e.g., *Oenococcus oeni*). Malolactic fermentation contributes to an olfactory change in the base wine, as it increases the concentration of diacetyl, acetoin, volatile acids, diethyl succinate, ethyl acetate, *n*-propanol, 2-butanol, hexanol, ethyl lactate, and 2,3-butanediol. These compounds decrease the perception of fruity aromas, giving rise to compounds that impart desirable aromas to Champagne, including notes of butter, toast, nuts, and lactic (Sereni *et al.* 2020). This practice was introduced in Champagne production to prevent the spontaneous malolactic fermentation during the second bottle fermentation, which could increase the volatile acidity of sparkling wine, causing a defect. The selection of yeast strains that becomes prevalent in relation to bacteria, as well as the use of sulfur dioxide, prevents the spontaneous occurrence of malolactic fermentation in the bottle. Therefore, the inoculation of bacteria to perform the malolactic fermentation in the base wine is no longer justified, since it increases the time and cost of producing sparkling wines (Jeandet *et al.* 1997).

The use of indigenous yeasts for the production of sparkling wine is a trend in this area. Studies have been focused on the use of these yeasts in still wines, but data on native yeasts that support the conditions of a second fermentation (CO_2, low pH, alcohol content) are incipient. This approach could consolidate the production of differentiated sparkling wines, with a strong impact from its terroir. This is an important tool for the valorization of a product, especially for emerging wine-growing regions, such as places in Australia, Brazil, New Zealand, South Africa, among others, which have appeared on the international market due to the good quality of their sparkling wines.

Volatile compounds formed from fermentation include mainly esters, fatty acids, alcohols, and carbonyl compounds, among others. Figure 5.3 shows the main pathways that encompass the formation of these classes of compounds. The formation and concentration of volatile compounds during fermentation depend on the production method, yeast strain, fermentation conditions, and the characteristics of the product to be obtained, as shown in Figure 5.3a. Furthermore, winemaking can follow different guidelines according to local wine legislation and take into account the oenologist expertise (Moreno-Arribas and Polo 2009).

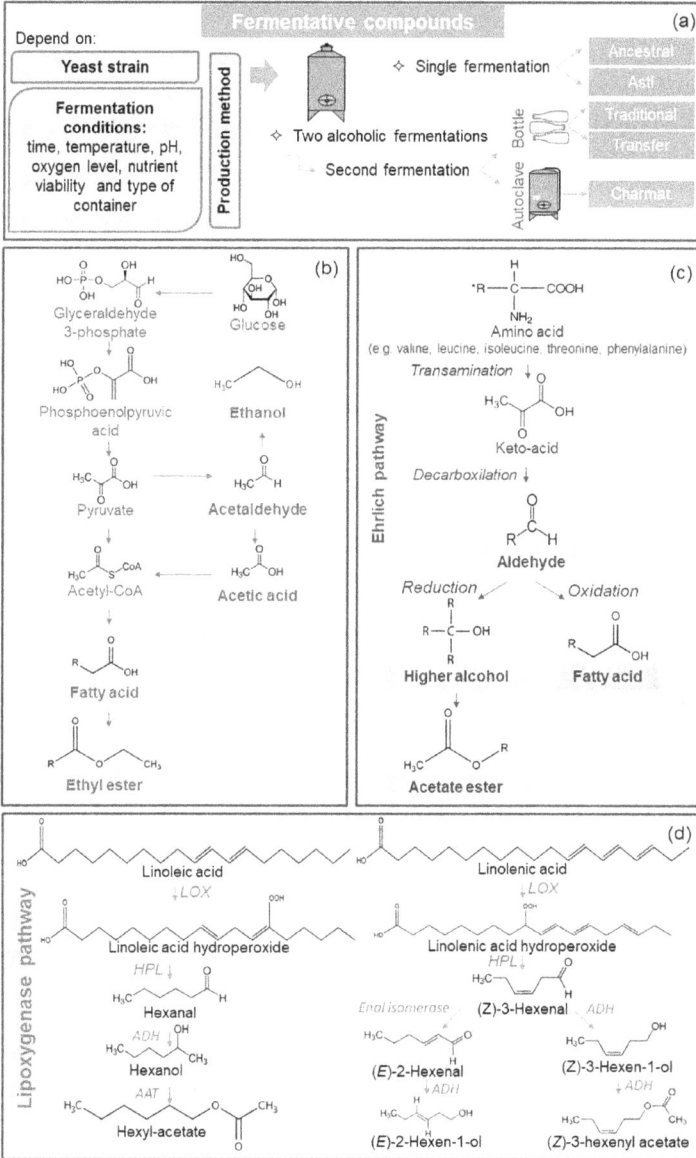

Figure 5.3   Sparkling Fermentative Compounds. (a) Factors That Drive the Formation of Fermentative Compounds. (b) Pathway for the Formation of Ethanol, Acetaldehyde, Acetic Acid, Fatty Acids, and Ethyl Esters. (c) Ehrlich Pathway That Results in the Formation of Aldehydes, Higher Alcohols, Fatty Acids, and Acetate Esters. (d) Lipoxygenase (LOX) Pathway Giving Rise to C6 Compounds by Hydroperoxide Lyases (HPL), Alcohol Acetyltransferase (AAT), and Alcohol Dehydrogenase (ADH).

5.3.2.1 Esters

Esters are the most important fermentative aroma class for sparkling wines. Although some of these compounds are below their perception threshold, the synergistic effect among different esters contributes to the fruity and floral aroma of the sparkling wines (Sumby, Grbin, and Jiranek 2010).

Esters are formed through the reaction between an alcohol and an acid catalyzed by enzymes (stearases and lipases) produced by yeasts (Hu *et al.* 2018). The main groups of esters found in sparkling wines are acetate esters and ethyl esters. Ethyl esters result from the reaction between ethanol and a fatty acid (Figure 5.3b). Acetate esters are formed from acetic acid and a higher alcohol derived from the amino acid metabolism (Ehrlich pathway, Figure 5.3c) (Sumby, Grbin, and Jiranek 2010). The production of esters is highly influenced by factors such as fermentation temperature, nutrient viability for yeasts, pH, presence of unsaturated fatty acids, and oxygen level (Styger, Prior, and Bauer 2011). Table 5.3 shows the esters commonly identified in sparkling wines, regardless of the method of elaboration, with their respective descriptor odors and odor thresholds.

Diethyl succinate, ethyl lactate, and ethyl isovalerate were identified as aging markers of sparkling wines produced by the Traditional method using País grapes from Chile (Table 5.1). However, the fruity and floral nuances of these sparkling wines reside in a few high-impact aromatic compounds, such as ethyl isobutyrate, isoamyl acetate, ethyl hexanoate, and diethyl succinate (Ubeda *et al.* 2019). Other esters as ethyl hexadecanoate, hexyl acetate, 2-phenylethyl acetate, and isoamyl acetate were identified as markers of Cava produced using a blend (50% Macabeo, 30% Xarel-lo, and 20% Parellada) from the Penedés grape-growing area (Spain) (Martínez-García *et al.* 2021). The fruity characteristic of Brazilian Moscatel sparkling wines can be attributed mainly to ethyl octanoate, ethyl decanoate, and hexyl acetate, whose concentrations increased during fermentation (Soares *et al.* 2015). Acetate esters of monoterpenes, as citronellyl, neryl, and geranyl, were reported for the first time in sparkling wines by Soares *et al.* (2015). The reaction between a monoterpene alcohol (citronellol, nerolidol, and geraniol, respectively) and acetic acid results in the formation of monoterpene acetates in the Moscatel sparkling wines, which can give floral notes to the aroma. However, there have been only a few studies on this type of sparkling wines, and much more is needed.

5.3.2.2 Fatty Acids

Some fatty acids present in the sparkling wine are considered volatile or semi-volatile. Acetic acid is the most abundant volatile acid, which is derived from sugar metabolism (Figure 5.3b), and, when detected in concentrations above $1.2 \, g \, L^{-1}$, it has a negative effect for sparkling wine aroma. Semivolatile acids

TABLE 5.3 ESTERS FOUND IN SPARKLING WINES REGARDLESS OF THE METHOD OF ELABORATION.

Compounds	Odor Descriptor	Odor Threshold (µg L)[*]
Isoamyl acetate	Banana, fruity, sweet	160[1]
Hexyl acetate	Apple, cherry, pear, floral	670[1]
Phenyl ethyl acetate	Rose	250[1]
Ethyl butanoate	Pineapple, apple	400[1]
Ethyl 2-methyl butanoate	Strawberry	1[2]
Ethyl hexanoate	Green apple, strawberry	8[1]
Ethyl octanoate	Sweet, fruity, pear	580[1]
Ethyl decanoate	Floral	500[1]
Diethyl succinate	Fruity, floral	1200[1]
Ethyl lactate	Fruity, buttery	150,000[1]
Ethyl isovalerate	Apple, sweet	1[2]
Ethyl isobutyrate	Fruity, strawberry, lemon	15[2]
2-Phenyl ethyl acetate	Honey, roses, flowery	250[2]

[*]Odor threshold determined in 10–12% (v/v) ethanol obtained from [1]Peinado *et al.* 2004 and [2]Guth 1997.

(odor described as fatty) including hexanoic, octanoic, and decanoic acid are precursors of ethyl esters such as ethyl hexanoate, ethyl octanoate, and ethyl decanoate, respectively. These esters play a fundamental role in the aroma of sparkling wines, contributing with notes of fruity aroma, regardless of the method of preparation (Ubeda *et al.* 2016; Hu *et al.* 2018).

Sparkling wines elaborated by methods that include two fermentations have an increased concentration of ethyl esters since the metabolism of medium chain fatty acids is favored in the second fermentation (C6–C10) (Caliari *et al.* 2015; González-Jiménez *et al.* 2020; Ubeda *et al.* 2016). However, for the metabolic route of ethyl esters formation from fatty acids to evolve properly during fermentation, it is important that several factors regarding yeasts are taken into account, such as nutritional status, oxygen availability, and temperature, among others (Waterhouse, Sacks, and Jeffery 2016).

5.3.2.3 Higher Alcohols

As esters and fatty acids, higher alcohols (alcohols having more than two carbon atoms) are also important fermentative compounds for the sparkling wine aroma (Styger, Prior, and Bauer 2011). The main biosynthesis pathway of higher alcohols is associated with the amino acid catabolism via the Ehrlich pathway (Figure 5.3c), which includes a series of transamination, decarboxylation, oxidation, and reduction reactions catalyzed by different enzymes produced by yeasts (Cameleyre *et al.* 2015). Isobutanol (ethereal), 1-propanol (alcoholic), 3-methyl-1-butanol (fermented), 2-methyl-1-butanol (roasted), and 2-phenyl ethanol (floral) are among the higher alcohols most often found in sparkling wines (the contribution to the aroma of each alcohol was mentioned in parentheses) (Coelho *et al.* 2009; Caliari *et al.* 2015; Welke *et al.* 2014).

The second fermentation carried out in the Traditional, Charmat, and Transfer methods (Figure 5.3a) favors the formation of higher alcohols (Caliari *et al.* 2015; Ubeda *et al.* 2016; Di Gianvito *et al.* 2018; Ubeda *et al.* 2019; Welke *et al.* 2014) The concentrations of these compounds tend to remain constant even after the end of fermentations and during autolysis, due to complex balances of intracellular synthesis and extracellular adsorption-desorption processes during lees aging (Martínez-García *et al.* 2017).

5.3.2.4 Carbonyl Compounds

Carbonyl compounds (ketones and aldehydes) have less impact than esters, fatty acids, and alcohols in regard to the fermentative aroma of sparkling wines. These compounds come from the metabolism of sugars or amino acids. Despite being produced during fermentation, they are also associated with the maturation stage of the sparkling wines and therefore can be important for postfermentative aromas (Torrens *et al.* 2010).

Acetaldehyde is the major aldehyde formed from the metabolization of sugars by yeasts (Figure 5.3b). It is important to mention that the metabolism of amino acids such as alanine also contributes to the formation of this compound. In addition, it can also be a result of the pathway of production of acetic acid from ethanol by bacteria. When detected in concentrations above 100 mg L^{-1}, acetaldehyde can present notes of rotten apple, which are associated with high levels of oxidation, and therefore it is a defect for sparkling wines (Waterhouse, Sacks, and Jeffery 2016).

Acetaldehyde can bind with SO_2, reducing the antimicrobial properties of SO_2 and making itself odorless (Coetzee, Buica, and du Toit 2018). It can also play an important sensory role on the color of rose and red sparkling wines, as it accelerates polymerization between anthocyanins and catechins or tannins, increasing the intensity and stability of color (Liu and Pilone 2000).

Ketones have not been frequently identified in sparkling wines. Diacetyl (2,3-butanedione) may be produced mainly during malolactic fermentation, which is rarely performed in sparkling wine production. The reduction of diacetyl results in acetoin, which can be reduced to 2,3-butanediol, giving pleasant buttery notes (Martínez-García *et al.* 2020; Waterhouse, Sacks, and

Jeffery 2016). However, 2,3-butanediol has not been found at concentrations above its odor threshold (Ubeda *et al.* 2019; Martínez-García *et al.* 2020).

5.3.2.5 C6 Compounds

C6 compounds including the alcohols hexanol, 3-hexenol, 2-hexenol and mainly the acetate ester called hexyl acetate have been reported in sparkling wines (Ubeda *et al.* 2019; Torrens *et al.* 2010; Carlin *et al.* 2016), as shown in Table 5.1. The unsaturated fatty acids, linoleic and linolenic acids, are the precursors of these compounds via the lipoxygenase pathway (LOX), as presented in Figure 5.3c. LOX and hydroperoxide lyases (HPL) are activated by the rupture of grape cells during the winemaking process, particularly during the pressing and mashing of grapes. These reactions give rise to a C6 aldehyde (hexanal, (*Z*)-3-hexenal or (*E*)-2-hexenal, followed by the respective alcohols (hexanol, (*Z*)-3-hexen-1-ol, and (*E*)-2-hexen-1-ol), which are characterized by a grassy and herbaceous aroma. Among these C6 alcohols, hexanol has been reported as a possible marker of sparkling wines (Table 5.1) produced by the Traditional (Ubeda *et al.* 2019; López de Lerma *et al.* 2018; Martínez-García *et al.* 2017; Carlin *et al.* 2016) and Asti methods (Soares *et al.* 2015). This alcohol is the precursor of hexyl acetate (fruity notes), which has been found to be a marker of aging of Cava sparkling wines (Traditional method, Denomination of Origin status from Spain) (Martínez-García *et al.* 2021; Martínez-García *et al.* 2017; Torrens *et al.* 2010; Riu-Aumatell *et al.* 2006).

5.3.3 Postfermentative Compounds

Postfermentative aromas are formed mainly during aging due to yeast autolysis. Furan compounds are formed in the postfermentative stage, including autolysis and storage. These compounds are derived from sugar degradation, giving desirable aroma notes described mainly as bread, caramel, almond, toasted, among others. Compounds such as furfural (fruity, caramel), 5-methyl furfural (fruity, caramel), 2-methyl-3-furan thiol (toasted), 2-furfuryl thiol (dried fruit), and acetyl furan (balsamic) were responsible for the complexity of postfermentative aromas in Cavas and positively influence the quality of these sparkling wines (Torrens *et al.* 2010).

Furan compounds are proposed as aging markers for both still and sparkling wines. Sparkling wines produced by the Traditional method undergo biological aging in contact with lees in anaerobic conditions, which favors the accumulation of furan compounds. 5-Hydroxymethyl-2-furfural (5-HMF) was identified as an aging marker in Cava (Martínez-García *et al.* 2021) and Champagne (Serra-Cayuela *et al.* 2013).

5.4 THE ROLE OF AUTOLYSIS ON VOLATILE PROFILE

After completing the fermentative steps, changes in the chemical composition of the sparkling wines associated with autolysis can occur. Autolysis occurs mainly in sparkling wines produced by the Traditional and Transfer methods, in which greater complexity and structure are sought (Kemp *et al.* 2015). Greater aromatic complexity due to reactions such as esterification, hydrolysis, reduction, and oxidation of the compounds present in the sparkling wine can contribute to the formation of volatile compounds that attribute notes such as toast, yeast, butter, dried fruits, and nuts (Torrens *et al.* 2010). Wine regions with Protected Designation of Origin for sparkling wines have distinct periods of time for autolysis such as Champagne and Cava (minimum 15 and 9 months, respectively) (Europe Commission 2008).

Autolysis allows the release of glucan, mannoproteins, proteins, amino acids, peptides, nucleotides, nucleosides, and lipids to the sparkling wine, which influence foaming characteristics, mouthfeel, and flavor, as shown in Figure 5.4a (Kemp *et al.* 2015). General changes that occur in the volatile profile due to yeast autolysis are as follows:

- Glucanases produced during autolysis are responsible for the hydrolysis of the glycosidic bond of volatile varietal compounds, such as terpenes and norisoprenoids, releasing them to the free form (Coelho *et al.* 2009; López de Lerma *et al.* 2018; Muñoz-Redondo *et al.* 2020; Torchio *et al.* 2012). Two C13-norisoprenoids have been identified as the main indicators of the autolysis process: 1,1,6-trimethyl-1,2-dihydro naphthalene (TDN) (Korenika *et al.* 2020; Muñoz-Redondo *et al.* 2020; López de Lerma *et al.* 2018; Torrens *et al.* 2010; Riu-Aumatell *et al.* 2006; Welke *et al.* 2014) and vitispirane (Ubeda *et al.* 2019; López de Lerma *et al.* 2018; Welke *et al.* 2014; Torrens *et al.* 2010; Riu-Aumatell *et al.* 2006). Two monoterpenes are among the most frequently reported in sparkling wines that undergo autolysis: α-terpineol and linalool.
- Esterases released by yeast degradation during autolysis lead to hydrolysis of esters such as hexyl acetate and 2-phenylethyl acetate (Martínez-García *et al.* 2021; Martínez-García *et al.* 2017; Torrens *et al.* 2010; Riu-Aumatell *et al.* 2006).
- Succinic acid esterification gives rise to diethyl succinate. It is important to mention that this ester is also formed during fermentation and in autolysis its concentration is increased (Korenika *et al.* 2020; Ubeda *et al.* 2016; Torrens *et al.* 2010; Riu-Aumatell *et al.* 2006).
- Sorption of volatile hydrophobic compounds (with log P values above 4, e.g., ethyl 9-decenoate, ethyl decanoate, and 2-methylbutyl octanoate) on the yeast lees. Depending on the time of autolysis, these compounds may be subsequently desorbed into sparkling wine (Gallardo-Chacón *et al.* 2010).

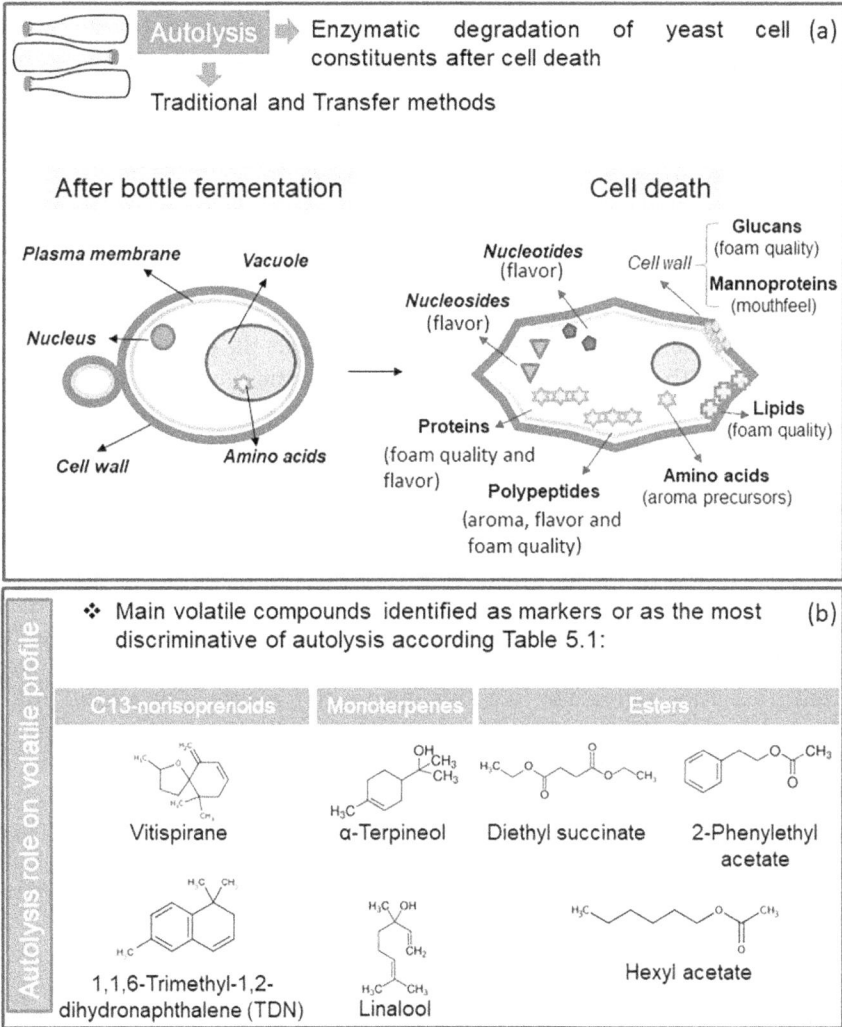

Figure 5.4 Autolysis Process That Occurs in the Traditional and Transfer methods (autolysis Can Occur in the Ancestral Method According to the Criteria Established by Each Region), Which Results in (a) the Release of Important Compounds to Sparkling Wine Quality, Including (b) Volatile Compounds.

It is important to note that the formation of volatile compounds due to autolysis also depends on the grape variety, yeast strain, fermentation conditions, among other factors. The stress conditions in which the yeast develops during the second fermentation in the bottle (low pH, presence of carbon dioxide and ethanol) and its autolytic capacity are key factors for the volatile profile of sparkling wines produced by the Traditional method (Di Gianvito *et al.* 2018).

In the Traditional method, once the aging time is brought to an end, riddling (*remuage*) takes place by manually turning the bottle around one-eighth of a turn every day for around 15 days. The purpose is to produce sediment facilitating the lees removal. The riddling stage entails an intensive and time-consuming workforce and its improvement awakens the interest of winemakers. Automated riddling has been implemented in large wineries with machines called gyropalettes` that hold 504 bottles per cage, reducing the riddling time to 2–4 days (Jeandet *et al.* 2000). In another approach, the developments in the use of immobilized yeasts may also shorten both the time and the expenses of this process (Moreno-García *et al.* 2018). During second fermentation and aging, the immobilized yeasts remain agglomerated and readily flocculate, unlike free yeasts, which are dispersed inside the bottle and sediment slowly (Costa *et al.* 2018). Yeast immobilization consists in the physical confinement of yeasts usually by using an external support containing calcium alginate (Costa *et al.* 2018), calcium alginate–chitosan (Benucci *et al.* 2019), oak chips, or cellulose powder (Berbegal *et al.* 2019), which are recommended due to their easy preparation and high degree of yeast adsorption. No significant differences in the volatile profile have been reported between wines produced with immobilized yeast using alginate or those obtained with free cells (Costa *et al.* 2018). However, as calcium ions may alter the foam characteristics (López de Lerma *et al.* 2018), yeast biocapsules have been proposed by the coimmobilization of *S. cerevisiae* in filamentous fungi with Generally Recognized as Safe (GRAS) status (such as *Penicillium chrysogenum*). Studies focused on applying biocapsules to the sparkling wine elaboration demonstrated that this immobilization system does not negatively affect their aroma quality and foamability (Puig-Pujol *et al.* 2013; López de Lerma *et al.* 2018).

Due to the importance of autolysis for the particular characteristics of wines, some technologies have been studied to accelerate this phenomenon, including high pressure homogenization (HPH) (Comuzzo *et al.* 2017), pulsed electric field (PEF) (Barba *et al.* 2015), ultrasound (Liu *et al.* 2016), and microwave (Gnoinski *et al.* 2021). The use of HPH to accelerate yeast autolysis in Chardonnay sparkling wines resulted in the decrease of medium- and long-chain esters such as ethyl isovalerate (ester of 3-methylbutyric acid), ethyl octanoate, ethyl tetradecanoate, and ethyl-11-hexadecenoate, which may be related to fruity aroma. Further research is needed to study the effects of HPH on the sensory attributes and foaming ability of sparkling wines (Patrignani *et al.* 2013). The effect of other technologies (PEF, ultrasound, and microwave) on the volatile profile of sparkling wines has not yet been studied. In addition, it is important to be aware that in many parts of the world, sparkling wines are produced by small wineries that may not be willing to invest in these types of technologies.

5.5 FINAL CONSIDERATIONS AND PERSPECTIVES

In the production of sparkling wines by the Traditional, Charmat, Transfer, Ancestral, and Asti methods, several stages/factors influence the expression of varietal, fermentative, and postfermentative volatile compounds. Regardless of the production method, fermentation plays a fundamental role in releasing the glycosylated form of varietal compounds (terpenes and C13-norisoprenoids). Several classes of volatile compounds are formed during fermentation including esters, alcohols, fatty acids, aldehydes, and ketones. The formation of these compounds is related to terroir, grape variety, yeast strain, and fermentation conditions, among other factors. The scientific literature has been dedicated mainly to the study of sparkling wines produced by the Traditional method, especially considering the role of autolysis in the volatile profile. 1,1,6-Trimethyl-1,2-dihydronaphthalene (TDN) and vitispirane have been identified as the main markers of the autolysis of sparkling wines produced with different grape cultivars in various locations around the world.

Although the production of sparkling wines is an ancient art, scientific literature is poor in this area, and many aspects still need to be studied. Wide varieties of styles are available in the market as a result of different methods of preparation and the introduction of distinct grape varieties. These varieties of products have attracted younger consumers (aged between 19 and 25 years), who prefer softer and lighter sparkling wines. This scenario stimulated the production of this type of sparkling wine in large volumes, mainly by the Charmat and Asti methods, which have been even more poorly studied.

REFERENCES

Alessandrini, M., F. Gaiotti, N. Belfiore, F. Matarese, C. D'Onofrio, and D. Tomasi. 2017. Influence of vineyard altitude on glera grape ripening (vitis vinifera L.): Effects on aroma evolution and wine sensory profile. *Journal of the Science of Food and Agriculture* 97 (9):2695–2705. doi:10.1002/jsfa.8093.

Alexandre, H., and M. Guilloux-Benatier. 2006. Yeast autolysis in sparkling wine—a review. *Australian Journal of Grape and Wine Research* 12 (2):119–127. doi:10.1111/j.1755-0238.2006.tb00051.x.

Alfonzo, A., N. Francesca, V. Mercurio *et al.* 2020. Use of grape racemes from grillo cultivar to increase the acidity level of sparkling base wines produced with different saccharomyces cerevisiae strains. *Yeast* 37 (9–10):475–486. doi:10.1002/yea.3505.

ASTM International. 2012. *ASTM, Designation: E1432 – 19. Standard Practice for Defining and Calculating Individual and Group Sensory Thresholds from Forced-Choice Data Sets of Intermediate Size.* https://standards.globalspec.com/std/13494708/astm-e1432-19.

Barba, F. J., O. Parniakov, S. A. Pereira *et al.* 2015. Current applications and new opportunities for the use of pulsed electric fields in food science and industry. *Food Research International* 77 (November):773–798. doi:10.1016/j.foodres.2015.09.015.

Benucci, I., M. Cerreti, D. Maresca, G. Mauriello, and M. Esti. 2019. Yeast cells in double layer calcium alginate—chitosan microcapsules for sparkling wine production. *Food Chemistry* 300 (January). Elsevier:125174. doi:10.1016/j.foodchem.2019.125174.

Berbegal, C., L. Polo, M. J. García-Esparza, V. Lizama, S. Ferrer, and I. Pardo. 2019. Immobilisation of yeasts on oak chips or cellulose powder for use in bottle-fermented sparkling wine. *Food Microbiology* 78 (December 2017). Elsevier Ltd: 25–37. doi:10.1016/j.fm.2018.09.016.

Buxaderas, S., and E. López-Tamames. 2012. Sparkling wines: Features and trends from tradition. *Advances in Food and Nutrition Research*: 66. doi:10.1016/B978-0-12-394597-6.00001-X.

Caliari, V., V. M. Burin, J. P. Rosier, and M. T. Bordignon-Luiz. 2014. Aromatic profile of brazilian sparkling wines produced with classical and innovative grape varieties. *Food Research International* 62. Elsevier Ltd:965–973. doi:10.1016/j.foodres.2014.05.013.

Caliari, V., C. P. Panceri, J. P. Rosier, and M. T. Bordignon-Luiz. 2015. Effect of the traditional, charmat and asti method production on the volatile composition of moscato giallo sparkling wines. *LWT—Food Science and Technology* 61 (2). Elsevier Ltd:393–400. doi:10.1016/j.lwt.2014.11.039.

Cameleyre, M., G. Lytra, S. Tempere, and J. C. Barbe. 2015. Olfactory impact of higher alcohols on red wine fruity ester aroma expression in model solution. *Journal of Agricultural and Food Chemistry* 63 (44):9777–9788. doi:10.1021/acs.jafc.5b03489.

Carlin, S., U. Vrhovsek, P. Franceschi *et al.* 2016. Regional features of Northern Italian sparkling wines, identified using solid-phase micro extraction and comprehensive two-dimensional gas chromatography coupled with time-of-flight mass spectrometry. *Food Chemistry:* 208. Elsevier Ltd:68–80. doi:10.1016/j.foodchem.2016.03.112.

Carrau, F. M., E. Boido, and E. Dellacassa. 2008. Terpenoids in grapes and wines: Origin and micrometabolism during the vinification process. *Natural Product Communications* 3 (4). doi:10.1177/1934578X0800300419.

Coelho, E., M. A. Coimbra, J. M. F. Nogueira, and S. M. Rocha. 2009. Quantification approach for assessment of sparkling wine volatiles from different soils, ripening stages, and varieties by stir bar sorptive extraction with liquid desorption. *Analytica Chimica Acta* 635 (2):214–221. doi:10.1016/j.aca.2009.01.013.

Coetzee, C, A. Buica, and W. J. du Toit. 2018. The use of SO2 to bind acetaldehyde in wine: Sensory implications. *South African Journal of Enology and Viticulture* 39 (2):157–162. doi:10.21548/39-2-3156.

Comuzzo, P., S. Calligaris, L. Iacumin, F. Ginaldi, S. Voce, and R. Zironi. 2017. Application of multi-pass high pressure homogenization under variable temperature regimes to induce autolysis of wine yeasts. *Food Chemistry* 224 (June):105–113. doi:10.1016/j.foodchem.2016.12.038.

Costa, G., K. P. Nicolli, J. E. Welke, V. Manfroi, and C. A. Zini. 2018. Volatile profile of sparkling wines produced with the addition of mannoproteins or lees before second fermentation performed with free and immobilized yeasts. *Journal of the Brazilian Chemical Society* 29 (9):1866–1875. doi:10.21577/0103-5053.20180062.

Croce, G. B. 2004. *Della Eccellenza e Diversità de i Vini, Che Nella Montagna Di Torino Si Fanno, e Del Modo Di Farli*. Torino: Edizioni SEB27.

Di Gianvito, P., G. Perpetuini, F. Tittarelli *et al.* 2018. Impact of saccharomyces cerevisiae strains on traditional sparkling wines production. *Food Research International* 109 (April). Elsevier:552–560. doi:10.1016/j.foodres.2018.04.070.

Dziadas, M., and H. H. Jeleń. 2010. Analysis of terpenes in white wines using SPE-SPME-GC/MS approach. *Analytica Chimica Acta* 677 (1):43–49. doi:10.1016/j.aca.2010.06.035.

Europe Commission (EC). 2008. *Council Regulation (EC) N° 479/2008 of 29 April 2008 on the Common Organisation of the Market in Wine*. Official Journal European Union. https://eur-lex.europa.eu/legal-content/EN/TXT/?uri=CELEX%3A32008R0479.

Gallardo-Chacón, J. J., S. Vichi, E. López-Tamames, and S. Buxaderas. 2010. Changes in the sorption of diverse volatiles by saccharomyces cerevisiae lees during sparkling wine aging. *Journal of Agricultural and Food Chemistry* 58 (23):12426–12430. doi:10.1021/jf103086e.

Ganss, S., F. Kirsch, P. Winterhalter, U. Fischer, and H. G. Schmarr. 2011. Aroma changes due to second fermentation and glycosylated precursors in chardonnay and riesling sparkling wines. *Journal of Agricultural and Food Chemistry* 59 (6):2524–2533. doi:10.1021/jf103628g.

Gnoinski, G. B., S. A. Schmidt, D. C. Close, K. Goemann, T. L. Pinfold, and F. L. Kerslake. 2021. Novel methods to manipulate autolysis in sparkling wine: Effects on yeast. *Molecules* 26 (2):387. doi:10.3390/molecules26020387.

González-Jiménez, M. del C., J. Moreno-García, T. García-Martínez *et al.* 2020. Differential analysis of proteins involved in ester metabolism in two saccharomyces cerevisiae strains during the second fermentation in sparkling wine elaboration. *Microorganisms* 8 (3). MDPI AG. doi:10.3390/microorganisms8030403.

González-Royo, E., O. Pascual, N. Kontoudakis *et al.* 2015. Oenological consequences of sequential inoculation with non-saccharomyces yeasts (torulaspora delbrueckii or metschnikowia pulcherrima) and saccharomyces cerevisiae in base wine for sparkling wine production. *European Food Research and Technology* 240 (5):999–1012. doi:10.1007/s00217-014-2404-8.

Guth, H. 1997. Identification of character impact odorants of different white wine varieties. *Journal of Agricultural and Food Chemistry* 45 (8):3022–3026. doi:10.1021/jf9608433.

Hu, K., G. J. Jin, W. C. Mei, T. Li, and Y. S. Tao. 2018. Increase of medium-chain fatty acid ethyl ester content in mixed H. Uvarum/S. Cerevisiae fermentation leads to wine fruity aroma enhancement. *Food Chemistry* 239 (January). Elsevier Ltd:495–501. doi:10.1016/j.foodchem.2017.06.151.

Ilc, T., D. Werck-Reichhart, and N. Navrot. 2016. Meta-analysis of the core aroma components of grape and wine aroma. *Frontiers in Plant Science* 7 (September). doi:10.3389/fpls.2016.01472.

International Organization for Standardization. 2002. *ISO 13301. Sensory Analysis—Methodology—General Guidance for Measuring Odour, Flavour and Taste Detection Thresholds by a Three-Alternative Forced-Choice (3-AFC) Procedure*. https://www.iso.org/standard/36791.html.

International Organization for Standardization. 2008. *ISO 5492:2008 — Sensory Analysis—Vocabulary*. https://www.iso.org/standard/38051.html#:~:text=ISO%20 5492%3A2008%20applies%20to,4)%20terminology%20relating%20to%20methods.

Ivit, N. N., I. Loira, A. Morata, S. Benito, F. Palomero, and J. A. Suárez-Lepe. 2018. Making natural sparkling wines with non-saccharomyces yeasts. *European Food Research and Technology* 244 (5). Springer Berlin Heidelberg:925–935. doi:10.1007/ s00217-017-3015-y.

Jackson, R. S. 2020. *Wine Science*, 5th ed. Amsterdam: Elsevier. doi:10.1016/ C2017-0-04224-6.

Jeandet, P., J. C. Lenfant, M. Caillet, G. Liger-Belair, Y. Vasserot, and R. Marchal. 2000. Contribution to the study of the remuage procedure in champagne wine production: Identification of voltigeurs (particles in suspension in the wine). *American Journal of Enology and Viticulture* 51 (4):418, LP-419. www.ajevonline.org/ content/51/4/418.abstract.

Jeandet, P., Y. Vasserot, G. Liger-Belair, and R. Marchal Marchal. 1997. Sparkling wine production. Edited by Joshi V. K. *Trends in Food Science & Technology* 8 (9). Asiatech Publishers Inc:315. doi:10.1016/s0924-2244(97)80273-7.

Jones, J. E., F. L. Kerslake, D. C. Close, and R. G. Dambergs. 2014. Viticulture for sparkling wine production: A review. *American Journal of Enology and Viticulture* 65 (4):407– 416. doi:10.5344/ajev.2014.13099.

Kemp, B., H. Alexandre, B. Robillard, and R. Marchal. 2015. Effect of production phase on bottle-fermented sparkling wine quality. *Journal of Agricultural and Food Chemistry* 63 (1). American Chemical Society:19–38. doi:10.1021/jf504268u.

Korenika, A. M. J., D. Preiner, I. Tomaz, and A. Jeromel. 2020. Volatile profile characterization of croatian commercial sparkling wines. *Molecules* 25 (18). doi:10.3390/ molecules25184349.

Lin, J., M. Massonnet, and D. Cantu. 2019. The genetic basis of grape and wine aroma. *Horticulture Research* 6 (1):81. doi:10.1038/s41438-019-0163-1.

Liu, L., I. Loira, A. Morata, J. A. Suárez-Lepe, M. C. González, and D. Rauhut. 2016. Shortening the ageing on lees process in wines by using ultrasound and microwave treatments both combined with stirring and abrasion techniques. *European Food Research and Technology* 242 (4):559–569. doi:10.1007/s00217-015-2566-z.

Liu, S., and G. J. Pilone. 2000. An overview of formation and roles of acetaldehyde in winemaking with emphasis on microbiological implications. *International Journal of Food Science and Technology* 35:49–61.

López de Lerma, N., R. A. Peinado, A. Puig-Pujol, J. C. Mauricio, J. Moreno, and T. García-Martínez. 2018. Influence of two yeast strains in free, bioimmobilized or immobilized with alginate forms on the aromatic profile of long aged sparkling wines. *Food Chemistry* 250 (January). Elsevier:22–29. doi:10.1016/j.foodchem.2018.01.036.

Marcon, A. R., L. V. Schwarz, S. V. Dutra *et al.* 2018. Contribution of a Brazilian Torulaspora Delbrueckii isolate and a commercial saccharomyces cerevisiae to the aroma profile and sensory characteristics of moscato branco wines. *Australian Journal of Grape and Wine Research* 24 (4):461–468. doi:10.1111/ajgw.12347.

Martínez-García, R., T. García-Martínez, A. Puig-Pujol, J. C. Mauricio, and J. Moreno. 2017. Changes in sparkling wine aroma during the second fermentation under CO2 pressure in sealed bottle. *Food Chemistry* 237:1030–1040. doi:10.1016/j. foodchem.2017.06.066.

Martínez-García, R., J. Moreno, A. Bellincontro *et al.* 2021. Using an electronic nose and volatilome analysis to differentiate sparkling wines obtained under different conditions of temperature, ageing time and yeast formats. *Food Chemistry* 334 (April 2020). Elsevier:127574. doi:10.1016/j.foodchem.2020.127574.

Martínez-García, R., Y. Roldán-Romero, J. Moreno, A. Puig-Pujol, J. C. Mauricio, and T. García-Martínez. 2020. Use of a flor yeast strain for the second fermentation of sparkling wines: Effect of endogenous CO2 over-pressure on the volatilome. *Food Chemistry* 308 (March). Elsevier Ltd. doi:10.1016/j.foodchem.2019.125555.

Mateo, J. J. and M. Jiménez. 2000. Monoterpenes in grape juice and wines. *Journal of Chromatography A* 881 (1–2):557–567. doi:10.1016/S0021-9673(99)01342-4.

McKay, M., F. F. Bauer, V. Panzeri, and A. Buica. 2020. Investigation of olfactory inter-actions of low levels of five off-flavour causing compounds in a red wine matrix. *Food Research International* 128 (November 2019). Elsevier:108878. doi:10.1016/j.foodres.2019.108878.

Montella, M. M. 2012. Marketing del cultural heritage territoriale e musei di impresa: Un caso di analisi. *Mercati e Competitività* 4:33–51. doi:10.3280/MC2012-004004.

Moreno-Arribas, M. V., and M. C. Polo. 2009. *Wine Chemistry and Biochemistry*. New York: Springer-Verlag.

Moreno-García, J., T. García-Martínez, J. C. Mauricio, and J. Moreno. 2018. Yeast immobili-zation systems for alcoholic wine fermentations: Actual trends and future perspec-tives. *Frontiers in Microbiology* 9 (February). doi:10.3389/fmicb.2018.00241.

Moreno-García, J., and R. Peinado. 2012. *Enological Chemistry*. San Diego, CA: Elsevier. doi:10.1016/C2011-0-69661-9.

Muñoz-Redondo, J. M., M. J. Ruiz-Moreno, B. Puertas, E. Cantos-Villar, and J. M. Moreno-Rojas. 2020. Multivariate optimization of headspace solid-phase microextraction coupled to gas chromatography-mass spectrometry for the analysis of terpenoids in sparkling wines. *Talanta* 208 (October 2019). Elsevier:120483. doi:10.1016/j.talanta.2019.120483.

OIV. International Organisation of Vine and Wine. 2020. *The Global Sparkling Wine Market OIV Focus*. International Organisation of Vine and Wine. https://www.oiv.int/public/medias/7291/oiv-sparkling-focus-2020.pdf.

Patrignani, F., M. Ndagijimana, P. Vernocchi *et al.* 2013. High-pressure homogenization to modify yeast performance for sparkling wine production according to traditional methods. *American Journal of Enology and Viticulture* 64 (2):258–267. doi:10.5344/ajev.2012.12096.

Peinado, R. A., J. Moreno, J. E. Bueno, J. A. Moreno, and J. C. Mauricio. 2004. Comparative study of aromatic compounds in two young white wines subjected to pre-fermentative cry-omaceration. *Food Chemistry* 84 (4): 585–590. doi:10.1016/S0308-8146(03)00282-6.

Pozo-Bayón, M. A., P. J. Martín-Álvarez, M. V. Moreno-Arribas, I. Andujar-Ortiz, and E. Pueyo. 2010. Impact of using Trepat and Monastrell red grape varieties on the volatile and nitrogen composition during the manufacture of rosé Cava spar-kling wines. *LWT – Food Science and Technology* 43 (10): 1526–1532. doi:10.1016/j.lwt.2010.05.023.

Puig-Pujol, A., E. Bertran, T. García-Martínez, F. Capdevila, S. Mínguez, and J. C. Mauricio. 2013. Application of a new organic yeast immobilization method for sparkling wine production. *American Journal of Enology and Viticulture* 64 (3):386–394. doi:10.5344/ajev.2013.13031.

Ribéreau-Gayon, P., Y. Glories, A. Maujean, and D. Dubourdieu. 2006. *Handbook of Enology*, 2nd ed. Chichester: John Wiley & Sons, Ltd. doi:10.1002/0470010398.

Riu-Aumatell, M., J. Bosch-Fusté, E. López-Tamames, and S. Buxaderas. 2006. Development of volatile compounds of cava (Spanish sparkling wine) during long ageing time in contact with lees. *Food Chemistry* 95 (2):237–242. doi:10.1016/j.foodchem.2005.01.029.

Ruiz-Moreno, M. J., J. M. Muñoz-Redondo, F. J. Cuevas, A. Marrufo-Curtido, J. M. León, P. Ramírez, and J. M. Moreno-Rojas. 2017. The influence of pre-fermentative maceration and ageing factors on ester profile and marker determination of Pedro Ximenez sparkling wines. *Food Chemistry* 230:697–704. doi:10.1016/j.foodchem.2017.03.048.

Sereni, A., Q. Phan, J. Osborne, and E. Tomasino. 2020. Impact of the timing and temperature of malolactic fermentation on the aroma composition and mouthfeel properties of chardonnay wine. *Foods* 9 (6):802. doi:10.3390/foods9060802.

Serra-Cayuela, A., M. Castellari, J. Bosch-Fusté, M. Riu-Aumatell, S. Buxaderas, and E. López-Tamames. 2013. Identification of 5-hydroxymethyl-2-furfural (5-HMF) in cava sparkling wines by LC-DAD-MS/MS and NMR spectrometry. *Food Chemistry* 141 (4). Elsevier Ltd:3373–3380. doi:10.1016/j.foodchem.2013.05.158.

Soares, R. D., J. E. Welke, K. P. Nicolli *et al.* 2015. Monitoring the evolution of volatile compounds using gas chromatography during the stages of production of moscatel sparkling wine. *Food Chemistry* 183. Elsevier Ltd:291–304. doi:10.1016/j.foodchem.2015.03.013.

Styger, G., D. Jacobson, and F. F. Bauer. 2011. Identifying genes that impact on aroma profiles produced by saccharomyces cerevisiae and the production of higher alcohols. *Applied Microbiology and Biotechnology* 91 (3):713–730. doi:10.1007/s00253-011-3237-z.

Styger, G., B. Prior, and F. F. Bauer. 2011. Wine flavor and aroma. *Journal of Industrial Microbiology & Biotechnology* 38 (9):1145–1159. doi:10.1007/s10295-011-1018-4.

Sumby, K. M., P. R. Grbin, and V. Jiranek. 2010. Microbial modulation of aromatic esters in wine: Current knowledge and future prospects. *Food Chemistry* 121 (1):1–16. doi:10.1016/j.foodchem.2009.12.004.

Sun, L., B. Zhu, X. Zhang *et al.* 2020. The accumulation profiles of terpene metabolites in three muscat table grape cultivars through HS-SPME-GCMS. *Scientific Data* 7 (1):5. doi:10.1038/s41597-019-0321-1.

Tominaga, T., G. Guimbertau, and D. Dubourdieu. 2003. Role of certain volatile thiols in the bouquet of aged champagne wines. *Journal of Agricultural and Food Chemistry* 51 (4):1016–1020. doi:10.1021/jf020755k.

Torchio, F., S. R. Segade, V. Gerbi *et al.* 2012. Changes in varietal volatile composition during shelf-life of two types of aromatic red sweet brachetto sparkling wines. *Food Research International* 48 (2). Elsevier Ltd:491–498. doi:10.1016/j.foodres.2012.04.014.

Torrens, J., M. Riu-Aumatell, S. Vichi, E. López-Tamames, and S. Buxaderas. 2010. Assessment of volatile and sensory profiles between base and sparkling wines. *Journal of Agricultural and Food Chemistry* 58 (4):2455–2461. doi:10.1021/jf9035518.

Tufariello, M., S. Pati, L. D'Amico, G. Bleve, I. Losito, and F. Grieco. 2019. Quantitative issues related to the headspace-SPME-GC/MS analysis of volatile compounds in wines: The case of maresco sparkling wine. *LWT* 108 (March). Elsevier:268–276. doi:10.1016/j.lwt.2019.03.063.

Ubeda, C., R. M. Callejón, A. M. Troncoso, A. Peña-Neira, and M. L. Morales. 2016. Volatile profile characterisation of chilean sparkling wines produced by traditional and charmat methods via sequential stir bar sorptive extraction. *Food Chemistry* 207. Elsevier Ltd:261–271. doi:10.1016/j.foodchem.2016.03.117.

Ubeda, C., I. Kania-Zelada, R. del Barrio-Galán, M. Medel-Maraboli, M. Gil, and Á. Peña-Neira. 2019. Study of the changes in volatile compounds, aroma and sensory attributes during the production process of sparkling wine by traditional method. *Food Research International* 119 (June 2018). Elsevier:554–563. doi:10.1016/j.foodres.2018.10.032.

Velázquez, R., E. Zamora, M. L. Álvarez, and M. Ramírez. 2019. Using torulaspora delbrueckii killer yeasts in the elaboration of base wine and traditional sparkling wine. *International Journal of Food Microbiology* 289 (April 2018). Elsevier:134–144. doi:10.1016/j.ijfoodmicro.2018.09.010.

Voce, S., D. Škrab, U. Vrhovsek, F. Battistutta, P. Comuzzo, and P. Sivilotti. 2019. Compositional characterization of commercial sparkling wines from Cv. Ribolla Gialla produced in Friuli Venezia Giulia. *European Food Research and Technology* 245 (10). Springer Berlin Heidelberg:2279–2292. doi:10.1007/s00217-019-03334-9.

Waterhouse, A. L., G. L. Sacks, and D. W. Jeffery. 2016. *Understanding Wine Chemistry*. Chichester: John Wiley & Sons, Ltd. doi:10.1002/9781118730720.

Welke, J. E., K. C. Hernandes, K. P. Nicolli, J. A. Barbará, A. C. Telles Biasoto, and C. A. Zini. 2021. Role of gas chromatography and olfactometry to understand the wine aroma: Achievements denoted by multidimensional analysis. *Journal of Separation Science* 44 (1):135–168. doi:10.1002/jssc.202000813.

Welke, J. E., M. Zanus, M. Lazzarotto, F. H. Pulgati, and C. A. Zini. 2014. Main differences between volatiles of sparkling and base wines accessed through comprehensive two dimensional gas chromatography with time-of-flight mass spectrometric detection and chemometric tools. *Food Chemistry* 164. Elsevier Ltd:427–437. doi:10.1016/j.foodchem.2014.05.025.

Yang, Y., G. J. Jin, X. J. Wang, C. L. Kong, J. B. Liu, and Y. S. Tao. 2019. Chemical profiles and aroma contribution of terpene compounds in Meili (Vitis Vinifera L.) grape and wine. *Food Chemistry* 284 (January). Elsevier:155–161. doi:10.1016/j.foodchem.2019.01.106.

Yuan, F., and M. C. Qian. 2016. Development of C13-norisoprenoids, carotenoids and other volatile compounds in vitis vinifera L. Cv. Pinot Noir grapes. *Food Chemistry* 192. Elsevier Ltd:633–641. doi:10.1016/j.foodchem.2015.07.050.

Chapter 6

Volatile Compounds Formation in Cider

Aline Alberti and Alessandro Nogueira

CONTENTS

6.1 INTRODUCTION

Cider, or hard cider, is a sparkling beverage obtained by the complete or partial alcoholic fermentation of apple musts from blends of dessert and/or industrial apples (Figure 6.1) (Santos *et al.* 2018).

Some 26,181.54 Th hectoliters were produced in 2019 in more than 35 countries and were consumed around the world (Figure 6.2). The leading markets are the United Kingdom, Spain, France, Germany, Ireland, and Poland, on the basis of the amount and the quality of the cider produced (Nogueira and Wosiacki 2016; AICV 2020). The global cider market is expected to grow by 6.1% between 2017 and 2023, due to the increase in the demand for gluten-free beverages and the increased preference for beverages with low alcohol content (1.2–8.5%). In addition, in the 2020s, cider is starting to be produced by microbreweries

DOI: 10.1201/9781003129462-8

Figure 6.1   Cider or Hard Cider: A Sparkling Alcoholic Beverage Made from Apples.

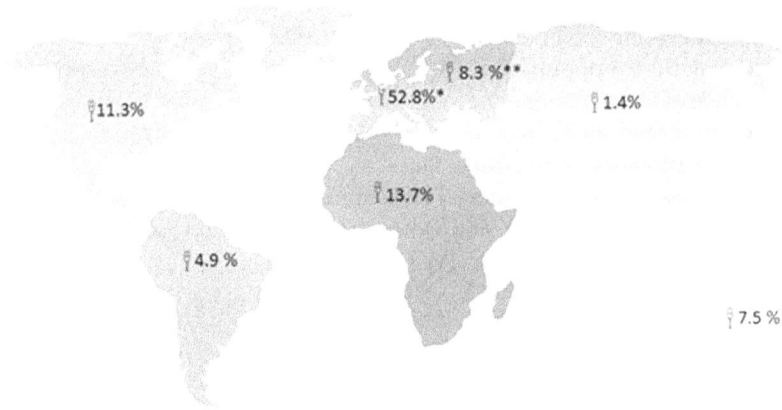

Figure 6.2   Distribution of Cider Consumption by Region of the World.

Note: *West Europe; **East Europe.

(where cider can also be served through a draft beer system, Figure 6.3) and multinational breweries (AICV 2020; Singh 2018).

Aroma is characterized as one of the most important indicators of cider quality (Fan, Xu, and Han 2011). The volatile compounds mixture that resulted in the apple beverage aroma has different reactive groups and chemical functions such as alcohols, aldehydes, ketones, carboxylic acids, esters, lactones, and terpenes (Le Quéré *et al.* 2006; Xu, Fan, and Qian 2007; Santos *et al.* 2015, 2016; Santos *et al.* 2018). Approximately 400 volatile compounds have already been identified and quantified in the apple, with about 20 to 40 of these playing

Figure 6.3 Cider Being Served from Draft Beer System.

a significant role in the aromatic profile of most cultivars (Abbas 2006; Qin, Petersen and Bredie 2018; Wei *et al.* 2019a; Braga *et al.* 2013). Many aromatic compounds from the apple are lost during the processing steps; however, most of the compounds responsible for the aroma of the cider are synthesized during fermentation (Williams and Tucknott 1978). The aromatic profile of ciders is extremely complex and has a direct relationship with the type and concentration of aromatic compounds, which can come from both fermentation and raw material (Villière *et al.* 2015; Rosend *et al.* 2019).

6.2 CIDER PROCESSING

More than one million tons of apples were used for cider processing in the world in 2018 (7.6% of the whole production) (AICV 2019). Apple varieties and processing technology can vary between countries, regions, and producers around the world. The fraction of each variety (acidic, sour, sweet, astringent, bitter and aromatic) in order to maintain a composition profile (Table 6.1) is reflected in the sensorial quality of the beverage (Lea and Drilleau 2003; Nogueira and Wosiacki 2016).

TABLE 6.1 CHEMICAL COMPOSITION OF CIDER.

Parameter	Minimum and Maximum Values
Total sugars (g/L)	0.0–75.4
Glucose (g/L)	0.0–36.6
Fructose (g/L)	0.0–33.0
Sucrose (g/L)	0.0–28.1
Total acidity (g/L)	0.5–7.0
pH	2.9–4.1
Ethanol (°GL)	2.1–8.5
nitrogen (mg/L)	15.7–310.3
Total phenols (mg/L)	230.7–2463.0

Sources: Qin, Petersen, and Bredie (2018); Alberti *et al* (2016); Nogueira *et al.* (2008); Li *et al.* (2020); Santos *et al.* (2018).

The first stage of cider processing (Figure 6.4) is the selection of the apple. Thus the apples are washed and sanitized by showering or immersion, and after that, they pass through a knife mill. In some countries during the milling process, potassium metabisulfite (70–150 mg/L) can be added in order to control the browning reaction, to delay a natural fermentation, and as an antimicrobial to prevent contamination since harvesting and processing occur in summer with high daily temperatures varying from 25°C to 35°C (Nogueira and Wosiacki 2016). In some industrial plants, enzymatic maceration (pectinases, cellulases, and hemicellulases) can be carried out before the extraction of apple must. Then the apple must extraction is carried out in a hydraulic press, and the liquid obtained is modified by adding industrial pectinolytic preparation in amounts according to the instructions of the manufacturer. Sulfur dioxide is generally added at this stage (75–300 mg/L SO_2) (Lea and Drilleau 2003). After depectinization and hydrocolloid flotation, within 6–10 hours, the must is racked, and the pomace sediment can be put through a second pressing, usually in a press filter. Other clarification methods can be performed depending on the cider manufacturer (Nogueira and Wosiacki 2016).

After this, the apple must can be naturally fermented due to the yeast present in the fruit, which is the custom in France where the processing temperature varies from 7°C to 11°C; however, in other countries, the apple must is usually inoculated with industrial dry active yeast. Alcoholic fermentation occurs in fermenters equipped with gas-escaping devices (airlock) to avoid the entry of oxygen but allowing excess CO_2 release. Fermentation can be continued until all the sugar is converted to ethanol and carbon dioxide, or it can be interrupted maintaining residual sugars in the final product. If the level of organic acids

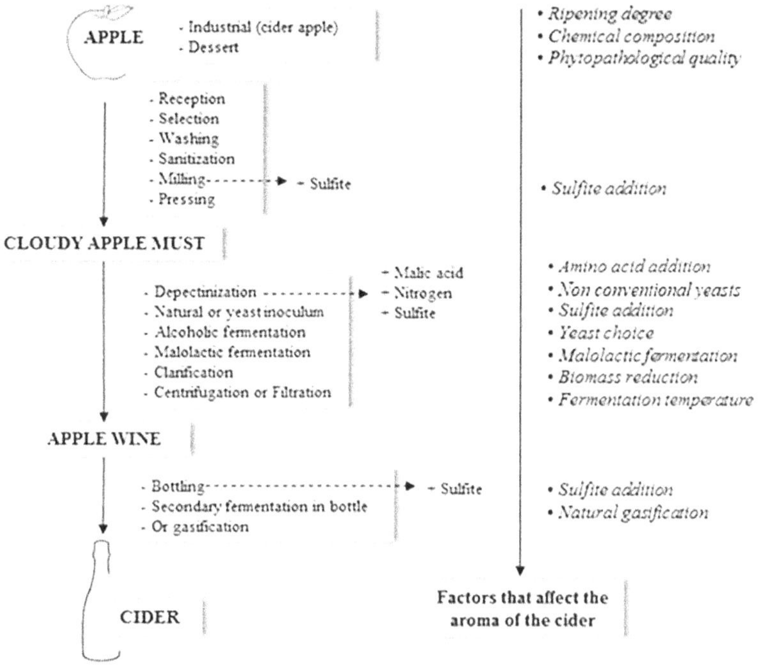

Figure 6.4   Cider-Making Process.

Note: "+" optional addition.

is high after the alcoholic fermentation, the product requires a resting period of around 20–35 days in order to support the transformation of malic to lactic acid (malolactic fermentation) and consequently to reduce the acid taste of the beverage. The deacidification of the beverage occurs through the action of indigenous lactic acid bacteria and can modify the aroma of the final cider (Li *et al.* 2020). In some countries, after the alcoholic fermentation, mixtures of the fermented must with or without sucrose are added in order to correct (variation of 40–100 g/L) the final sugar level. If necessary, the acidity is corrected with lactic or citric acid. The addition of conservatives normally conforms to potassium metabisulfite with 20–80 mg/L of free SO_2 and calcium sorbate at a concentration allowed by law (Wosiacki *et al.* 1997). Then the cider can be clarified by protein glue, generally with gelatine or casein, using bentonite as coadjuvant. The next step is stabilization by thermal treatment or conservant utilization, carbon dioxide addition at low temperature (artificial gasification), bottling in glass bottles at low temperature as well as in high-pressure resistant bottles with security cages in the stoppers, and finally labeling.

6.3 CIDER VOLATILE COMPOUNDS

Hundreds of volatile compounds, isolated from different yeast strains, that contribute to the aroma of the cider have been identified (Lea and Drilleau 2003, Pietrowski *et al.* 2012; Santos *et al.* 2015; Santos *et al.* 2018). The production of aroma in cider may be influenced by the environment, apple variety, ripening stage, processing, yeast strain, fermentation condition, and aging (Lea and Drilleau 2003; Braga *et al.* 2013; Rosend *et al.* 2019). The mains compounds that form the cider aroma complex are the esters, higher alcohols, fatty acids, aldehydes, ketones, terpenes, and lactones (Herrero, Garcia, and Diaz 2006, Qin, Petersen, and Bredie 2018). The volatile compounds 3-methyl-1-butanol, 2-phenyl ethanol, 1-hexanol, ethyl ethanoate, 2-phenyl ethyl ethanoate, ethyl hexanoate, ethyl octanoate, ethyl decanoate, ethyl 2-methyl butanoate, 3-methyl butyl ethanoate, ethyl butanoate, hexyl ethanoate, butanoic acid, hexanoic acid, octanoic acid and 2-methyl butanoic acid are considered responsible for the fruity aroma of cider (Le Quéré *et al.* 2006; Xu, Fan, and Qian 2007; Santos *et al.* 2015, 2016, 2018; Qin, Petersen, and Bredie 2018). The odors descriptors of major volatile compounds found in cider are described in Table 6.2.

TABLE 6.2   MAJOR VOLATILE COMPOUNDS DETECTED IN CIDERS AND THEIR RESPECTIVE ODOR DESCRIPTORS.

Volatile Compound	Odor Descriptors
Alcohols	
3-methyl 1-butanol	Alcohol, whisky, malt, burnt, fruity
2- phenyl ethanol	Floral, rose, lilac, honey
1-hexanol	Resin, floral, herbaceous, grass
Esters	
Ethyl acetate	Fruit, pineapple, vinegar, green
2-phenyl ethyl acetate	Rose, honey
3-methyl butyl acetate	Banana
Hexyl acetate	Fruit, herb, green apple, banana
Ethyl hexanoate	Apple peel, fruit, green apple, strawberry, floral
Ethyl octanoate	Fruit, fat, pineapple
Ethyl decanoate	Fruit, grape, rose
Ethyl lactate	Fruit, strawberry, floral
Ethyl-2-methyl butanoate	Fruity, apple
Ethyl butanoate	Apple, fruit, strawberry, pineapple, cognac
Acids	
Butanoic acid	Butter, cheese, floral
Hexanoic acid	Sweat, fatty, cheese, rancid

Volatile Compound	Odor Descriptors
Octanoic acid	Fatty, cheese, rancid
2-Methyl butanoic acid	Cheese, fermented
Aldehyde	
Acetaldehyde	Pungent, fruit
Ketone	
Acetoin	Butter, cream

Sources: Bingman *et al.* (2020); Simonato, Lorenzini, and Zapparoli (2021); Qin, Petersen, and Bredie (2018); Le Queré *et al.* (2006).

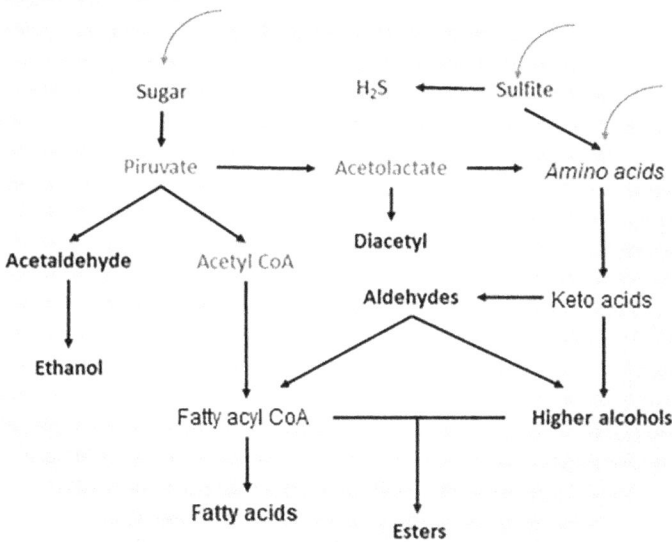

Figure 6.5   Volatile Compounds Formation During Cider Fermentation.

Sources: Abbas (2006), Henschke and Jiranek (1993).

Part of the aroma compounds from apples are lost in processing, and most of the aroma compounds in cider are synthesized during fermentation by yeast metabolism (Figure 6.5) (Williams and Tucknott 1978; Rosend *et al.* 2019; Peng *et al.* 2015; Qin, Petersen, and Bredie 2018).

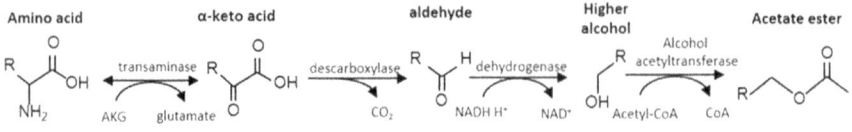

Figure 6.6   Catabolic Pathway of Amino Acid and the Formation of the Volatile Compounds.

Ehrlich's metabolic pathway is often involved in the formation of aromatic compounds, is a predominant pathway in yeast secondary metabolism, and is active mainly if amino acids are the only nitrogen source for yeast metabolism (Ye, Yue, and Yuan 2014; Li *et al.* 2015). The amino acids aspartate, asparagine, glutamate, and alanine, which are found in apples, are the main precursors of the volatile compounds in cider (Santos *et al.* 2015, 2016). The amino acid is transaminated and decarboxylated to the corresponding aldehyde, which is then reduced to higher alcohol. If there is an activity of the enzyme alcohol acetyltransferase, the alcohol can be partially transesterified to the acetate ester (Figure 6.6).

Esters are the most important aroma class, qualitatively, in cider. They are responsible for sweet and fruity odors. Some esters found in cider are present in apple must; nevertheless most of them are formed through the esterification of alcohols with fatty acids during fermentation (Xu, Zhao, and Wang 2006). In fermented fruit beverages, there is a wide variety of alcohols and acids, so the number of esters that can be formed is very variable. In these fermented beverages, the esters can originate in two different ways: by enzymatic esterification during the fermentation process or by chemical esterification during the aging period. In general, the amount of esters increases during the aging period (Ribéreau-Gayon *et al.* 2012).

The acetate esters are produced by the reaction of acetyl CoA with higher alcohols that are formed from amino acids or carbohydrates catabolism (Tufariello, Capone, and Siciliano 2012). Among esters, one of the most significant compounds that affect flavor in cider is ethyl ethanoate, produced from acetyl CoA and ethanol. At a proper concentration, ethyl ethanoate can contribute to fruit odor, even though it can negatively affect odor when levels are over 200 mg/L (Herrero, García, and Díaz 2006; Peng *et al.* 2015). Ethyl hexanoate, ethyl octanoate, ethyl decanoate, ethyl laurate, and ethyl lactate were also prominent among the esters.

According to the research of Peng *et al.* (2015), 20°C was considered the most suitable for cider making. At 20°C, the concentration of the ethyl acetate, isobutyl acetate, isopentyl acetate, ethyl caprylate, ethyl 4-hydroxybutanoate, isobutyl alcohol, isopentyl alcohol, 3-methylthio-1-propanol, and benzene ethanol are the highest and consequently the highest in acceptability in terms of sensory analysis. However, the authors emphasize that the results refer to the yeast strain *Saccharomyces cerevisiae* used in the study.

Higher alcohols are precursors of esters and are equally important for the aromatic sensory quality of apple fermented beverages (Li *et al*. 2015). As previously mentioned, these compounds are synthesized through the Ehrlich pathway in the presence of amino acids or sugars. 3-Methyl-butanol is derived from leucine and is considered the most quantitatively important alcohol in Chinese ciders. This alcohol was the compound identified in the highest concentration in fermented beverages made with apples from different ripening stages, and it was also found in the apple must by Braga *et al*. (2013). 1-Hexanol is considered a characteristic volatile compound of fruits, and it is synthesized during the enzymatic oxidation of linoleic acid (Fan, Xu, and Han 2011). The 2-methyl-1-propanol compound is synthesized from vanillin, while 2-methyl and 3-methyl-1-butanol are derived from isoleucine and leucine, respectively (Rosend *et al*. 2019).

Aldehydes are known for their contribution from apple-like to citrus-like to nutty in foods. Acetaldehyde or ethanal is commonly found in alcoholic beverages, and it is produced by the decarboxylation of pyruvate by yeast during alcoholic fermentation (Qin, Petersen, and Bredie 2018).

Ketones are formed by the condensation of fatty acids (Tufariello, Capone, and Siciliano 2012). Acetoin is detected in cider after malolactic fermentation, and it contributes positively to butter and cream notes of the cider aroma (Qin, Petersen, and Bredie 2018). Braga *et al*. (2013) also identified low levels of hexanone, 2-heptanone, and 2-octanone in ciders made with Gala cultivar at the ripe and senescent stage. Another important compound is diacetyl, which is synthesized from pyruvate by *Leuconostoc* species during malolactic fermentation and positively contributes to the buttery aroma in wines and ciders. Even at this stage, aromas that can be described as spicy and phenolic can be formed, especially when fruits are classified as acid-bitter. These compounds are the ethyl-phenol and catechol resulting from hydrolysis, decarboxylation, and reduction of p-coumaroylquinic and chlorogenic acid, respectively (Beech and Carr 1977; Bartowsky and Henschke 2004). Ciders of superior quality taste sweet and fruity and fragrant, while products with spicy, choking aromas and with acidic, astringent flavors are not well accepted.

The volatile acids can be formed during alcoholic fermentation. Acetic acid results from the oxidation of acetaldehyde, and its contents are related to yeast strain and the raw material used. Fatty acids are also found, especially butanoic acid and octanoic acid (Li *et al*. 2020). The latter one is found in high contents in cider from senescent Fuji apples. Octanoic acid is often found in Fuji apple wine (Braga *et al*. 2013; Rita *et al*. 2011). Fatty acids, formed enzymatically during fermentation, constitute another important group of aroma compounds that can contribute to fruity, cheese, fatty, and rancid notes (Qin, Petersen, and Bredie 2018).

Phenolic compounds are also contributors to cider aroma with animal, spicy, and smoky notes; however, these volatile compounds are related to certain olfactory defects. The ethyl phenols are originated from the enzymatic processes of yeast and aging (Lentz 2018). The conversion of phenolic compounds to volatile phenols can occur through the sequential activity of two enzymes, cinnamate decarboxylase and vinyl phenol reductase, which can be released by some yeasts of the genus *Saccharomyces* (Chatonnet *et al.* 1992). In addition, lactic acid bacteria, *Lactobacillus colloids*, and *Lactobacillus mali* are mainly responsible for the conversion of phenolic acids into 4-ethylphenol (stables, sweaty saddles, animal, and leather), 4-vinylguaiacol (clove, curry), and 4-ethyl guaiacol (smoky, spicy). The reaction can take place through two distinct pathways. In the first, phenolic acid is decarboxylated by the enzyme phenolic acid decarboxylase and reduced to ethylphenol by vinyl phenol reductase. The other way of conversion consists in the reduction of phenolic acid in a catalyzed reaction by phenolic acid reductase, followed by decarboxylation by the action of hydroxyphenyl propionic decarboxylase (Buron *et al.* 2011; Silva *et al.* 2011).

6.4 INFLUENCE OF PROCESSING ON THE VOLATILE PROFILE

6.4.1 Raw Material

The chemical composition of the apple must strongly influence the fermentation process. As the composition of the must can vary according to the variety of apples used, it is to be expected that the variety influences the volatile composition of the cider. According to Rosend *et al.* (2019) most of the higher alcohols, some esters (butyl acetate, ethyl-3-methyl butanoate, and isoamyl acetate), 3-methylbutanoic acid, 3-octanone, benzaldehyde, and phenylacetaldehyde, found in ciders, are influenced by the ripening stage of the apples. Braga *et al.* (2013) suggest using apple at the senescent stage to obtain a cider with higher volatile compounds content. Nevertheless, Villière *et al.* (2015) report that the process steps in cider processing have significantly more impact on cider aroma than apple variety when using a bitter blend of apples. A significant effect of apple blend, however, was observed in butyl acetate, ethyl-2- methylbutanoate, ethyl-3-methylbutanoate, butanoic acid, and 2-methyl-butanoic acid contents. The ripening stages of the apples used for juice and cider processing affect the whole chemical composition, as can be seen in Figure 6.7 (Alberti *et al.* 2016).

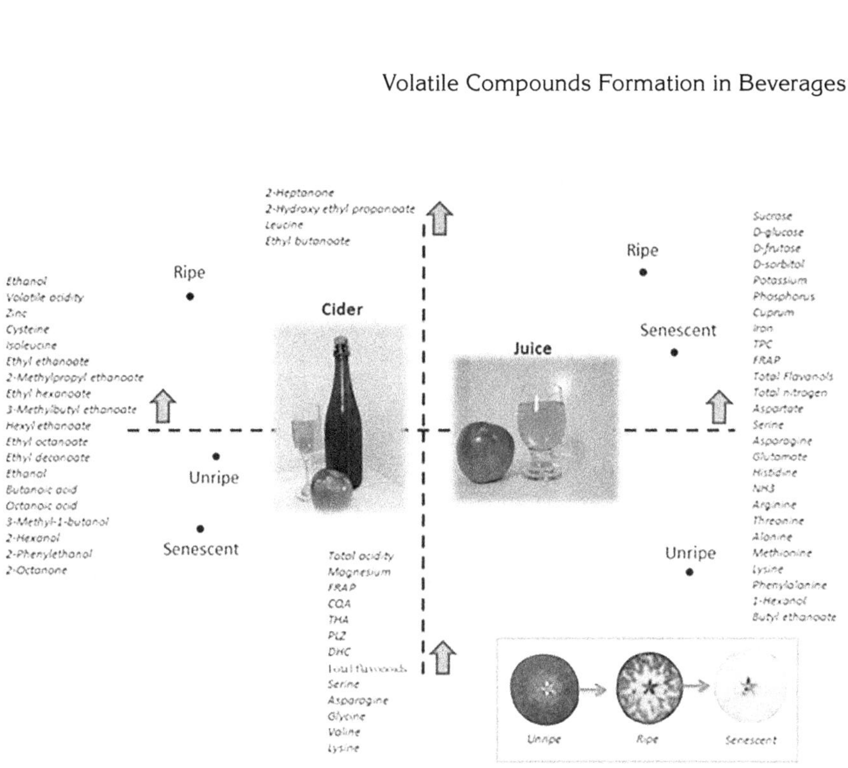

Figure 6.7   Effect of the Ripening Stage on the Chemical Composition of Apple Juices and Ciders.

Note: Upward arrows indicate an increase in concentration.

Source: Alberti *et al.* (2016).

Another parameter for the aromatic quality of the cider is the fungal contamination of the apples, mainly in the core that cannot be seen on the surface. Ciders made with apples with phytosanitary problems have worse sensory quality. The decrease in the content of aromatic compounds is probably related to a reduced precursor's availability (Simonato, Lorenzini, and Zapparoli 2021).

6.4.2 Clarification of the Apple Must

After the extraction, the clarification of apple must is carried out in order to remove compounds that may cause turbidity or that may negatively affect the cider sensory quality. The clarification steps can modify the volatile profile of the beverage and improve cider aroma (Le Quéré *et al.* 2006; Villière *et al.* 2015). The clarification not only eliminates the cloudy compounds but also can reduce the nitrogen compounds by adsorption (Valdés *et al.* 2011). When the must has low nitrogen content, yeast synthesizes more, higher alcohols, and consequently more esters are formed. Villière *et al.* (2015) observed that clarification by keeving methods enhances the content of the acetate esters in cider, which are known to contribute to the fruity aroma of alcoholic beverages. Dynamic keeving is a clarification method that consists of the addition of pectin methylesterase and calcium chloride. After gelation of the demethylated pectin, nitrogen gas is bubbled, and the pectinate gel floats; thus the clear juice can be racked off.

6.4.3 Alcoholic and Malolactic Fermentation

The fermentation process that involves the transformation of apple must into cider can involve three phases: oxidative, alcoholic, and malolactic (Braga *et al.* 2013, Nogueira *et al.* 2008). The oxidative phase involves nonconventional yeasts located on the surface of the apples. These microorganisms are the first to develop, and they can have high or low alcoholic fermentation activity and are responsible for the production of volatile compounds that will sensorially classify the ciders with fruity or floral notes (Pietrowski *et al.* 2012; Santos *et al.* 2018). Alcoholic fermentation starts after the oxidative phase in a natural fermentation process, but when yeasts of the genus *Saccharomyces* are inoculated, they inhibit the growth of oxidative yeasts.

The yeast strains used in the production of alcoholic beverages contribute to their volatile profile. Thus the selection of the strain with an aptitude for volatiles compounds formation is an essential step in the development of high-quality ciders. There are commercial yeasts specialized in the formation of aroma in ciders; they were often developed for sparkling wines but end up having a great result in ciders. Rosend *et al.* (2019) associated the compounds 2-methyl-1-propanol, ethyl propionate, ethyl-2-methyl butanoate, ethyl-3-methyl butanoate, butanoic acid, and 2-methylbutanoic acid with the yeast strain used for fermentation.

Although fermentation with pure cultures of *S. cerevisiae* provides advantages of easy control and homogeneous fermentation, the production of wines with these pure monocultures can result in a lack of complexity in the aromas. The aroma ends up being described as neutral or not typical. The inclusion of non-*Saccharomyces* yeasts as part of mixed cultures with *S. cerevisiae* has been suggested as a way to take advantage of spontaneous fermentation and improve the flavor characteristics of the final product (Rojas *et al.* 2003; Ciani, Beco, and Comitini 2006; Xu, Zhao, and Wang 2006; Wei *et al.* 2019a). Among them, yeast from the genus *Hanseniaspora* stands out for its high capacity to form aromatic compounds, especially the ability to form acetates, which are important aromatic contributors. As a result of these mixed fermentations, Xu, Fan, and Qian (2007) observed that the strains of *Hanseniaspora valbyensis* produce high levels of ethyl esters, that the strains of *S. cerevisiae* produce high amounts of alcohol, and that the result of these levels is a positive contribution to the final product. The major esters formed are ethyl acetate and phenyl ethyl acetate (considered to be the most influential on the aroma), and the alcohols are isoamyl alcohol and isobutyl. Within the *Hanseniaspora* strain, the species *H. uvarum* produces high levels of ethyl acetate, but, unlike other yeasts, this one produces acetoin in multiculture trials. This species does not cause an increase in volatile acidity in mixed cultures, does not maintain residual sugar, and uses less nitrogen in both mixed and sequential fermentations, indicating the absence of competition between them (Rojas *et al.* 2003; Ciani, Beco, and Comitini 2006).

In the past few years, non-*Saccharomyces* yeast strains aptitude has been extensively studied. Wei *et al.* (2019a) investigated the use of 163 non-*Saccharomyces* yeast strains and observed that *Candida zemplinina*, *Hanseniaspora vineae*, and *Torulaspora delbrueckii* produced more volatile compounds. Among the aromas identified in ciders, aldehydes, terpenes, and esters were the most prevalent common aromatic substances. In other work, the same authors observed that *P. kluyveri*, *H. vineae*, *H. uvarum*, and *T. quercuum* used in monocultures exhibited excellent performance in cider fermentation (Wei *et al.* 2019b). Fuji apple juice was the best substrate for *Pichia kluyveri* compared to the Gala and Red Star cultivars. The yeast previously mentioned contributes to the tropical fruit aroma of wines probably due to the presence of the 1-butanol, 3-methyl-, acetate; acetic acid, hexyl ester; and acetic acid, pentyl ester.

Other works analyzed sequential fermentation only with non-*Saccharomyces* yeasts. The simultaneous fermentation with *Hanseniaspora osmophila* and *Torulaspora quercuum* produced ciders with a temperate fruity aroma. 3-Methyl-1-butanol, ethyl 2-methylbutanoate, phenylethyl alcohol, β-phenethyl acetate, and β-damascenone were the major volatile compounds responsible for this sensorial description (Wei *et al.* 2020).

Malolactic fermentation is a secondary fermentation where malic acid is converted to lactic acid. This fermentation is conducted by lactic acid bacteria such as *Oenococcus*, *Lactobacillus*, *Leuconostoc*, and *Pediococcus*. *Oenococcus oeni* is the most common. Li *et al.* (2020) observed an increase in the higher alcohols (1-propanol, 2-methyl-1-propanol, 1-butanol, 1-hexanol, benzyl alcohol, and phenylethyl alcohol), fatty acids (hexanoic acid, decanoic acid, and 2-methylbutanoic acid), and terpenoids (alpha-farnesene, linalool, and citronellol) after malolactic fermentation in red-flesh apple cider. The authors also suggest that *O. oeni* can degrade diacetyl and release 2,3-butanediol. The effect of the malolactic fermentation seems to be affected by the bacteria strain (Sánchez *et al.* 2014). The fermentation with *Leuconostoc mesenteroides* subsp. *mesenteroides* can improve the levels of glycerin and benzyl ethanol, improving the floral aroma, besides the content of ethyl acetate, ethyl lactate, 3-hydroxy-2-butanone, butyrolactone, furfuryl alcohol, 4-hydroxy-butanone, dihydroxyacetone, isoamyl alcohol, and, 2,3-butanediol (Zhao *et al.* 2014).

6.4.3.1 Methods to Slow Down the Fermentation Rate

It is already known that slowing the fermentation rate increases the production of volatile compounds in wines and ciders (Nogueira *et al.* 2008; Le Quéré *et al.* 2006). The slow phase of fermentation is also intended to obtain a product with better sensorial quality and brings practical benefits for the producers, facilitating the organization of the treatments during the process. To reduce the speed of fermentation, some technological strategies are used, such as low temperatures, clarification by flotation, rackings, filtrations, and centrifugations (Lea and Drilleau 2003; Nogueira *et al.* 2008).

Low temperatures can reduce the activity of yeast cells. Temperatures below 5°C provide interesting results concerning aromas but require costly installations with thermal insulation and adequate cooling capacity (Drilleau 1991). The temperature range in the fermentation of the cider is 7–15°C, interfering in its speed; however, this is not the only factor (Lepage 2002).

The flotation method also affects fermentation kinetics as confirmed experimentally with a test without clarification (control), which lasts 20 days, compared to the same must clarified by the flotation method, which can reach 48 days (Lepage 2002). The elimination of yeast and bacteria in the *chapeau brun* promotes a nutritional depletion of the must in particular by reducing the concentration of nitrogen (Beech and Challinor 1951). In addition to lowering the speed of fermentation, this practice allows a clear fermentation, considered favorable to the formation of aroma constituents. When flotation is performed correctly, the fermentation stops naturally before all the sugars are consumed (Lea and Drilleau 2003; Nogueira *et al.* 2008). Traditionally, producers carry out successive rackings in order to remove the precipitated biomass, called sludge, reducing the nutritional potential of the medium by the same principle of flotation. However, when the initial fermentation rate is high, this causes a convection current in the system that prevents sedimentation. Currently, these two practices (flotation and racking) are rarely sufficient to slow fermentation. Therefore, flotation and racking are often ineffective, and thus it is necessary to compensate for this change in the raw material by centrifugation or filtration. In order to systematize the removal of the biomass, a clarification is performed (during the growth of the yeast) by filtration or centrifugation, after the consumption of 10 g.L^{-1} of sugars. Until this moment, the maximum population is not achieved, and growth can restart providing a slow fermentation rate (Le Quéré 1991; Nogueira *et al.* 2008).

Centrifugation is widely used to decrease the amount of biomass and the consequent speed of fermentation. Besides, biomass reduction can be performed at the beginning of the fermentation in order to reduce the transformation of sugars to ethanol. Thus it is possible to obtain a beverage that maintains a high sugar content and possibly with more microbiological stability (Lea and Drilleau 2003).

Although a fermentation conducted at a slower rate favors the aromatic quality of the cider, this process must be conducted carefully. A strong decrease in velocity during fermentation could lead to a sluggish or stuck fermentation. In addition, when the fermentation rate has dramatically reduced the content of acetates and linear fatty acid, ethyl esters in the cider can be smaller (Villière *et al.* 2015)

6.4.4 Sulfur Dioxide Addition

Sulfur dioxide (SO_2) addition in cider processing is a common practice. However, in wines and ciders, excessive sulfur dioxide must be avoided, not only for health reasons but also because it can cause sensory changes to neutralize aromas and contribute to the formation of undesirable sulfur compounds (Lea and Drilleau 2003; Guerrero and Cantos-Villar 2015). The losses of volatile compounds occur regardless of the moment of addition (crushed apple, apple must, or cider), concentration, and apple variety, and it can reach 97%. The synthesis of 3-methyl-1-butanol and its presence in the ciders are not affected by sulfur dioxide. On the other hand, acetaldehyde binds with sulfur dioxide, causing a reduction of up to 80% of the volatile compound. Another important volatile compound found in cider, the ethyl ethanoate also is reduced when the compound was added before fermentation (Santos *et al.* 2018). Therefore, the use of sulfur dioxide in the production of cider should be carried out with caution since it affects the composition of volatile compounds and consequently the sensory quality of the beverage.

6.5 CONCLUSION

This review indicates that aroma is recognized as one of the most important indicators of cider quality. The volatile compounds are synthesized during the biotransformation of apple juice into cider. This process can be divided into oxidative, alcoholic, and malolactic fermentations. Furthermore, the production of aroma in cider may be influenced by the environment, apple variety, ripening stage, processing, yeast strain, fermentation condition, and aging. The main compounds that form the cider aroma complex are the esters, higher alcohols, fatty acids, aldehydes, ketones, terpenes, and lactones. The volatile compounds 3-methyl-1-butanol, 2-phenyl ethanol, 1-hexanol, ethyl ethanoate, 2-phenyl ethyl ethanoate, ethyl hexanoate, ethyl octanoate, ethyl decanoate, ethyl 2-methyl butanoate, 3-methyl butyl ethanoate, ethyl butanoate, hexyl ethanoate, butanoic acid, hexanoic acid, octanoic acid, and 2-methyl butanoic acid are considered responsible for the fruity aroma of this beverage. The aromatic quality of cider continues to be a theme in several research centers in order to understand and standardize the cider's aromatic profile.

REFERENCES

Abbas, C. A. 2006. Production of antioxidants, aromas, colours, flavours, and vitamins by yeasts. In *Yeasts in Food and Beverages*, edited by A. Querol and G. H. Fleet, 285–334. Berlin Heidelberg: Springer-Verlag.

AICV. European Cider and Fruit Wine Association. 2019. *European Cider Trends 2019*. Accessed January 22. https://aicv.org/en/publications.

AICV. European Cider and Fruit Wine Association. 2020. *European Cider Trends 2020*. Accessed January 22. https://aicv.org/en/publications.

Alberti, A., T. P. M. Santos, A. A. F. Zielinski *et al.* 2016. Impact on chemical profile in apple juice and cider made from unripe, ripe and senescent dessert varieties. *LWT—Food Science and Technology* 65:436–443.

Bartowsky, E. J., and P. A. Henschke. 2004. The 'buttery' attribute of wine—diacetyl—desirability, spoilage and beyond. *International Journal of Food Microbiology* 96:235–252.

Beech, F. W., and S. W. Challinor. 1951. Maceration and defecation in cider-making I. Changes occurring in the pectin and nitrogen contents of apple juices. *Annual Report of Agricultural and Horticulture Research Station* 143–161.

Beech, F. W., and J. G. Carr. 1977. Cider and Perry. In *Economic Microbiology*, edited by H. A. Rose, 139–313. London: Academic Press.

Bingman, M. T., C. E. Stellick, J. P. Pelkey, J. M. Scott, and C. A. Cole. 2020. Monitoring cider aroma development throughout the fermentation process by headspace solid-phase microextraction (HS-SPME) gas chromatography-mass spectrometry (GC-MS) analysis. *Beverages* 6:40.

Braga, C. M., A. A. F. Zielinski, K. M. Silva *et al.* 2013. Classification of juices and fermented beverages made from unripe, ripe and senescent apples based on the aromatic profile using chemometrics. *Food Chemistry* 141:967–974.

Buron, N., M. Coton, C. Desmarais *et al.* 2011. Screening of representative cider yeasts and bacteria for volatile phenol-production ability. *Food Microbiology* 28:1243–1251.

Chatonnet, P., D. Dubourdieu, J. N. Boidron, and M. Pons. 1992. The origin of ethylphenols in wines. *Journal of the Science of Food and Agriculture* 60:165–178.

Ciani, M., L. Beco, and F. Comitini. 2006. Fermentation behaviour and metabolic interactions of multistarter wine yeast fermentations. *International Journal of Food Microbiology* 108:239–245.

Drilleau, J. F. 1991. Produits cidricoles. Quelques mots sur les composés phénoliques (tanins). *Technologie* 23:21–22.

Fan, W., Y. Xu, and Y. Han. 2011. Quantification of volatile compounds in Chinese ciders by stir bar sorptive extraction (SBSE) and gas chromatography-mass spectrometry (CG—MS). *Journal of the Institute of Brewing* 117:61–66.

Guerrero, R. F., and E. Cantos-Villar. 2015 Demonstrating the efficiency of sulphur dioxide replacements in wine: A parameter review. *Trends in Food Science & Technology* 42:27–43.

Henschke, P. A., and V. Jiranek. 1993. Yeasts-metabolism of nitrogen compounds. In *Wine Microbiology and Biotechnology*, edited by G. H. Fleet, 77–164. Chur, Switzerland: Harwood Academic Publishers.

Herrero, M., L. A. Garcia, and M. Diaz. 2006 Volatile compounds in cider: Inoculation time and fermentation temperature effects. *Journal of the Institute of Brewing* 112:210–214.

Lea, A. G. H., and J. F. Drilleau. 2003. Cidermaking. In *Fermented Beverage Production*, edited by A. G. H. Lea and J. R. Piggott, 59–87. New York: Klumer Academic, Plenum Publishers.

Lentz, M. 2018. The impact of simple phenolic compounds on beer aroma and flavor. *Fermentation* 4:20–33.

LePage, A. 2002. Cidre de variété Guillevic vers la maîtrise d´un cidre de qualité supérieure. Scienses et Techniques des Industries Agro-Alimentaire. In *Conservatoire National des Arts et Metiers,* 158. Paris.

Le Quéré, J. M. 1991. Fermentation lente du cidre: Pour une élaboration de qualité. *Pomme* 22:17–19.

Le Quéré, J. M., F. Husson, C. M. G. C. Renard, and J. Primault. 2006. French cider characterization by sensory, technological and chemical evaluations. *LWT—Food Science and Technology* 39:1033–1044.

Li, C. X., X. H. Zhao, W. F. Zuo, T. L. Zhang, Z. Y. Zhang, and X. S. Chen. 2020. The effects of simultaneous and sequential inoculation of yeast and autochthonous *Oenococcus oeni* on the chemical composition of red-fleshed apple cider. *LWT—Food Science and Technology* 124:109184.

Li, S., Y. Nie, Y. Ding, J. Zhao, and X. Tang. 2015. Effects of pure and mixed Koji cultures with Saccharomyces cerevisiae on apple homogenate cider fermentation. *Journal of Food Processing and Preservation* 39:2421–2430.

Nogueira, A., J. M. Le Quéré, P. Gestin, A. Michel, G. Wosiacki, and J. F. Drilleau. 2008. Slow fermentation in French cider processing due to partial biomass reduction. *Journal of the Institute of Brewing* 114:102–110.

Nogueira, A., and G. Wosiacki. 2016. Sidra. In *Bebidas Alcoólicas: Ciência e Tecnologia*, edited by W. Venturini-Filho, 184–212. São Paulo: Blucher.

Peng, B., F. Li, L. Cui, and Y. Guo. 2015. Effects of fermentation temperature on key aroma compounds and sensory properties of apple wine. *Journal of Food Science* 80:S2937–S2943.

Pietrowski, G. A. M., C. M. E. Santos, E. Sauer, G. Wosiacki, and A. Nogueira. 2012. Influence of fermentation with *Hanseniaspora* sp yeast on the volatile profile of fermented apple. *Journal of Agricultural and Food Chemistry* 60:9815–9821.

Qin, Z., M. A. Petersen, and W. L. P. Bredie. 2018. Flavor profiling of apple ciders from the UK and Scandinavian region. *Food Research International* 105:713–723.

Ribéreau-Gayon, P., D. Duboudieu, B. Donèche, and A. Lonvaud. 2012. *Traité d' oenologie*: *Vol. 1 of microbiologie du vin et vinifications*, 6th ed. Paris: La Vigne.

Rita, R., K. Zanda, K. Daina, and S. Dalija. 2011. Compositions of aroma compounds in fermented apple juice: Effect of apple variety, fermentation temperature and inoculated yeast concentration. *Procedia Food Science* 1:1709–1716.

Rojas, V., J. V. Gil, F. Piñaga, and P. Manzanares. 2003. Acetate ester formation in wine by mixed cultures in laboratory fermentations. *International Journal of Food Microbiology* 86:181–188.

Rosend, J., R. Kuldjärv, S. Rosenvald, and T. Paalme. 2019. The effects of apple variety, ripening stage, and yeast strain on the volatile composition of apple cider. *Heliyon* 5:e01953.

Sánchez, A., G. Revel, G. Antalick, M. Herrero, L. A. García, and M. Díaz. 2014. Influence of controlled inoculation of malolactic fermentation on the sensory properties of industrial cider. *Journal of Industrial Microbiology & Biotechnology* 41:853–867.

33333

2

2

2

2

Santos, C. E. M., A. Alberti, G. A. M. Pietrowski *et al.* 2016. Supplementation of amino acids in apple must for the standardization of volatile compounds in ciders. *Journal of the Institute of Brewing* 122:334–341.

Santos, C. M. E., G. A. M. Pietrowski, C. M. Braga *et al.* 2015. Apple amino acid profile and yeast strains in the formation of fusel alcohols and esters in cider production. *Journal of Food Science* 80:1170–1177.

Santos, T. P. M., A. Alberti, P. Judacewski, A. A. F. Zielinski, and A. Nogueira. 2018. Effect of sulphur dioxide concentration added at different processing stages on volatile composition of ciders. *Journal of the Institute of Brewing* 124:261–268.

Silva, I., F. M. Campos, T. Hogg, and J. A. Couto. 2011. Wine phenolic compounds influence the production of volatile phenols by wine-related lactic bacteria. *Journal of Applied Microbiology* 111:360–370.

Simonato, B., M. Lorenzini, and G. Zapparoli. 2021. Effects of post-harvest fungal infection of apples on chemical characteristics of cider. *LWT—Food Science and Technology* 138:110620.

Tufariello, M., S. Capone, and P. Siciliano. 2012. Volatile components of Negroamaro red wines produced in Apulian Salento area. *Food Chemistry* 132:2155–2164.

Singh, S. 2018. *Cider Market by Product, Distribution Channel and Packaging—Global Opportunity Analysis and Industry Forecast, 2017–2023.* Allied Market Research. https://www.alliedmarketresearch.com/cider-market.

Wei, J. P., S. Wang, Y. Zhang, Y. Yuan, and T. Yue, T. 2019a. Characterization and screening of non-*Saccharomyces* yeasts used to produce fragrant cider. *LWT—Food Science and Technology* 107:191–198.

Wei, J. P., Y. X. Zhang, Y. H. Yuan, L. Dai, and T. L. Yue. 2019b. Characteristic fruit wine production via reciprocal selection of juice and non-*Saccharomyces* species. *Food Microbiology* 79:66–74.

Wei, J. P., Y. Zhang, Y. Qiu *et al.* 2020. Chemical composition, sensorial properties, and aroma-active compounds of ciders fermented with *Hanseniaspora osmophila* and *Torulaspora quercuum* in co- and sequential fermentations. *Food Chemistry* 306:125623.

Williams, A. A., and O. G. Tucknott. 1978. The volatile aroma components of fermented ciders: Minor neutral components from the fermentation of sweet coppin apple juice. *Journal of the Science of Food and Agriculture* 29:381–397.

Wosiacki G, R. A. Cherubin, and D. S. Santos. 1997. Cider processing in Brazil. *Fruit Processing* 7:242–249.

Valdés, E., M. Vilanova, E. Sabio, and M. J. and Benalte. 2011. Clarifying agents effect on the nitrogen composition in must and wine during fermentation. *Food Chemistry* 125:430–437.

Villière, A., G. Arvisenet, R. Bauduin, J. M. Le Quéré, and S. Thierry. 2015. Influence of cider-making process parameters on the odourant volatile composition of hard ciders. *Journal of the Institute of Brewing* 121:95–105.

Xu, Y., W. Fan, and M. C. Qian. 2007. Characterization of aroma compounds in apple cider using solvent-assisted flavor evaporation and headspace solid-phase microextraction. *Journal of Agricultural and Food Chemistry* 55:3051–3057.

Xu, Y., G. A. Zhao, and L. P. Wang. 2006. Controlled formation of volatile components in cider making using a combination of *Saccharomyces cerevisiae* and *Hanseniaspora valbyensis* yeast species. *Journal Industrial of Microbiology and Biotechnology* 33:192–196.

Ye, M., T. Yue, and Y. Yuan. 2014. Changes in the profile of volatile compounds and amino acids during cider fermentation using dessert variety of apples. *European Food Research and Technology* 239:67–77.

Zhao, H., F. Zhou, P. Dziugan *et al.* 2014. Development of organic acids and volatile compounds in cider during malolactic fermentation. *Czech Journal of Food Sciences* 32:69–76.

Chapter 7

Volatile Compounds Formation in Kefir

Maria Gabriela da Cruz Pedrozo Miguel, Angélica
Cristina de Souza, Débora Mara de Jesus Cassimiro,
Disney Ribeiro Dias, and Rosane Freitas Schwan

CONTENTS

7.1 INTRODUCTION

Kefir is a fermented acidic beverage with low alcohol content, originally from the Northern Caucasian mountains, produced by adding kefir grains to milk or a sucrose solution with or without fruit extracts (Marshall and Cole 1985; Arslan 2015; Lynch *et al.* 2021). The term "kefir" originates from *kef,* a Turkish word, which can be translated to a good feeling for the sensation experienced after drinking it or promoting health claims. Kefir grains are small, complex, irregularly shaped, transparent, mucilaginous, or yellowish-white, measuring about 1–6 mm or sometimes up to 15 mm in diameter, with the appearance of miniature cauliflowers (Figure 7.1). These gelatinous grains are a symbiotic culture of bacteria and yeast

DOI: 10.1201/9781003129462-9

Figure 7.1   Macroscopic Structure of Kefir Grains: (a) milk kefir Grains; (b) water kefir Grains.

embedded in a resilient polysaccharide matrix, primarily composed of α-glucans in water kefir (WK) and kefiran in milk kefir (MK) (Lynch *et al.* 2021).

The commercial production and consumption of kefir beverages are widespread and popular in many countries, including the Caucasus Mountains of Russia, Europe, Asia, and South and North America (Plessas *et al.* 2017; Gut *et al.* 2021). As a wide microbial diversity is found in kefir grains, its adaptation in different substrates can be easier than fermentation using single-species starter cultures (Fiorda *et al.* 2017). Reports of kefir making with milk from sheep, goats, and cows, whey, coconut, soy milk, fruit pulp, and vegetable juice (Liu *et al.* 2002; Puerari *et al.* 2012; Hsieh *et al.* 2012; Cais-Sokolińska *et al.* 2015; Corona *et al.* 2016; Magalhães-Guedes *et al.* 2020; Wang *et al.* 2021). However, kefir made from cow's milk remains the most prevalent.

A diverse microbiota has been found from kefir grains using culture-dependent and culture-independent methods (Magalhães-Guedes *et al.* 2010; Miguel *et al.* 2010, 2011). The WK and MK are similar in their structure, associated microorganisms, and products formed during the fermentation process. Lactic acid bacteria (LAB) are the primary bacterial members of the complex grain community, particularly species of *Lactobacillus, Leuconostoc, Lactococcus, Streptococcus,* and *Pediococcus* (Simova *et al.* 2002; Witthuhn *et al.* 2005; Angulo *et al.* 1993; Takizawa *et al.* 1998; Garrote *et al.* 2001; Miguel *et al.* 2010, 2011; Dertli and Çon 2017). Acetic acid bacteria (AAB) have a secondary role, depending on the presence of oxygen such as *Acetobacter, Gluconobacter,* and *Glucanocetobacter* (Farnworth and Farnworth 2005; Garrote *et al.* 2001; Miguel *et al.* 2010, 2011; Marsh *et al.* 2013; Dertli and Çon 2017). Yeast dominant members are species belonging to *Candida, Saccharomyces, Kazachstania, Kluyveromyces, Pichia, Issatchenkia, Dekkera, Zygosaccharomyces,* and *Yarrowia* were reported to be found in kefir grains (Witthuhn and Schoeman 2004; Witthuhn *et al.* 2005; Garrote *et al.* 2001; Miguel *et al.* 2010, 2011, 2013; Dertli and Çon 2017).

Microbial interaction between yeast and bacteria during kefir fermentation processes produces substances that impact the flavor and quality of the final product. Lactic acid bacteria can produce specific metabolites, such as carboxylic acids and ketones associated with cheesy flavors, esters associated with fruity flavors, and 2,3-butanedione associated with buttery flavors and with acetic acid (Walsh *et al.* 2016). *Saccharomyces* and other fermentative yeasts are stimulated, such as covert acetaldehyde to ethanol by the action of the alcohol dehydrogenase enzyme, leading to high alcohol content. The yeast metabolites also seem to stimulate AAB growth that benefits from increased ethanol production to acetic acid metabolism, associated with vinegary flavors (Güzel-Seydim *et al.* 2000; Schwan *et al.* 2015; Walsh *et al.* 2016; Fiorda *et al.* 2017).

Although the formation of flavor and aroma of traditional kefir varies according to the substrate and microbial community of kefir grains (Beshkova *et al.* 2003), the rich chemical composition is usually observed in kefir-based beverages, including sugars (sucrose, glucose, and fructose), organic acids (lactic, acetic, citric, tartaric, butyric, malic, and propionic acids), alcohols (ethanol, hexanol, and glycerol), and esters (ethyl propionate, ethyl hexanoate, octanoate, and decanoate). Besides general descriptions and kefir characteristics, this chapter focuses on the main volatile compounds (VOCs) responsible for the distinctive flavors of MK and WK.

7.2 KEFIR PRODUCTION AND APPLICATIONS

Kefir can be produced by a continuous process where the kefir is taken out, and fresh milk is added. The traditional production method of the kefir beverage involves adding the kefir grains directly to the preferred substrate. The raw milk substrate must be boiled and cooled to 24–26°C. The pH value is decreased to 4.6 and inoculated with 2–10% (v/v) of grains, which should ferment for 18–24 hours at room temperature. The fermented milk is then separated from Kefir grains by filtration. The used grains can be dried at room temperature, kept at low temperature, and used again to inoculate a new substrate

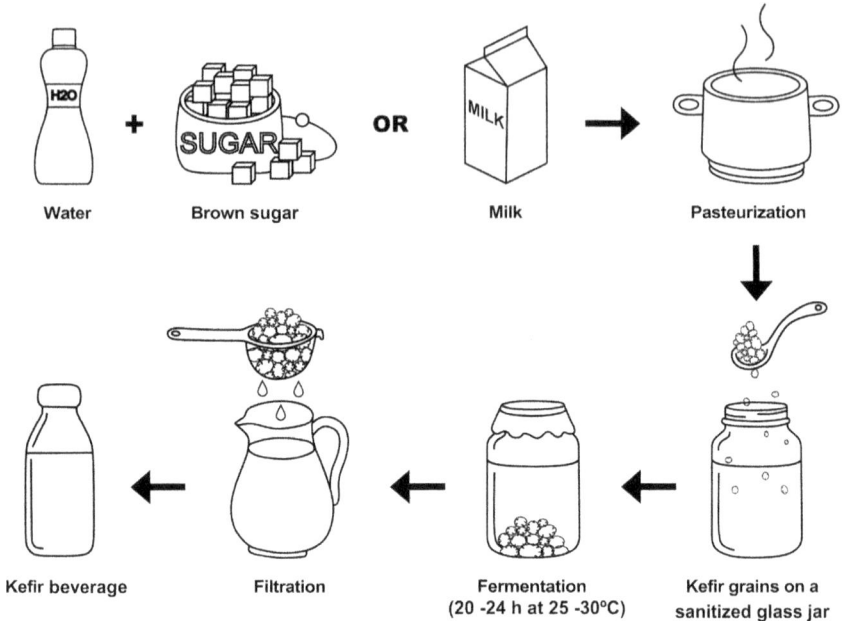

Figure 7.2   Traditional Flowchart of Milk and Water Kefir Beverage Homemade Production.

(Figure 7.2) (Güzel-Seydim *et al.* 2000; Miguel *et al.* 2010). Most of the studies are described by the traditional method, but some authors report Kefir beverage production by the Russian method and a large-scale industrial method. The Russian method involves a two-step fermentation, and the large-scale industrial method involves the use of pure kefir cultures rather than the kefir grain (Chunchom *et al.* 2017; Plessas *et al.* 2017; Liu *et al.* 2002). However, Dias and contributors (2012) reported problems with obtaining starter and stable cultures to maintain the final product quality. Garofalo and collaborators (2020) recently produced Kefir beverage by a back slopping method (fermented milk obtained from kefir grain fermentation). However, the results demonstrated that different kefir production methods affect the quality characteristics of the final product and that kefir beverages have been characterized in terms of their physicochemical and colorimetric parameters.

Kefir has been consumed as a potentially probiotic beverage and has shown several benefits besides being considered functional (Tu *et al.* 2019). Some properties can be mentioned such as antimicrobial (Rodrigues *et al.* 2005; Kim *et al.* 2016), anti-inflammatory (Rodrigues *et al.* 2005; Lee *et al.* 2007), healing (Rodrigues *et al.* 2005), antiallergic, cholesterol-lowering capability (Huang *et al.* 2013), and antioxidant activities (Zhang *et al.* 2017). Barbosa and contributors (2011) report that, due to the kefiran among the sugars, kefir also has anti-inflammatory and antimicrobial activities. Studies still report beneficial effects such as antitumorigenic and antistress properties and immunomodulatory and hypocholesterolemic functions in animal models (Meydani and Ha 2000; Rodrigues *et al.* 2005; Gaware *et al.* 2011).

Studies have been carried out for many years using kefir grains or beverage to produce and elaborate new kefir products. Plessas and contributors (2005) evaluated bread production using kefir grains and observed that the fermentation rate was reasonable and produced loaves of bread of good quality that was acceptable to tasters. Walnut is a crop of high economic interest to the food industry (Cui *et al.* 2013). The authors used kefir grains as the inoculum for the preparation of walnut milk beverages, showing positive results. Cais-Sokolinska and contributors (2015) reported that, in the development of kefir-based dairy products, the following factors play a crucial role: the chemical composition of the processed milk and the type and quantity of starter cultures. The same authors evaluated the production of kefir using goat and sheep milk and highlighted the changes that occur in volatile compounds. Generally, kefir is prepared using bovine milk. However, studies show that it can be prepared with other types of milk, such as buffalo, camel, sheep, and goat; rice, coconut, and soy beverages have also been used (Altay *et al.* 2013; Bourrie *et al.* 2016). Cheese whey and deproteinized cheese whey also have been used to produce kefir beverages (Magalhães-Guedes *et al.* 2011a, 2011b). The microorganisms present in the kefir grains were able to utilize lactose and produce volatile compounds similar to those obtained during traditional milk fermentation, which was reported by Magalhães-Guedes *et al.* (2011b).

A cocoa pulp–based kefir beverage was elaborated by (Puerari *et al.* 2012), and the final beverage met with great acceptance, the best of which might be for low concentrations of acidity/alcohol in the beverages. Another application for MK was the elaboration of an apple-based kefir vinegar (Viana *et al.* 2017). Kefir grains enabled the apple must to produce ethanol and acetic acid, as well as volatile alcohols and aldehydes, in vinegar-based kefir. Other types of milk to make the kefir beverage are also being attempted. Wang and colleagues (2021) used goat milk to produce a kefir beverage and evaluated its volatile compounds; kefir grains were also used to ferment a whey-based beverage and whey-based cheese by (Magalhães-Guedes *et al.* 2011a, 2011b), who showed that kefir grains contributed to the organoleptic characteristics of the final product. A new product was developed by Magalhães-Guedes *et al.* (2020): mango-based popsicles using MK grains; the authors reported that the new product allowed for the consumption of kefir biomass (by the ingestion of grains) after the fermentation. The development of arrowroot flour fermented by MK grains was shown by Souza *et al.* (2020), specifically that the fermentation process carried out allowed for the development and implementation of arrowroot flour fermented by kefir. According to the authors, the kefir grains developed in the arrowroot substrate, reduced the concentration of carbohydrates, produced organic acids, promoted an increase in phenolic compounds, and had high antioxidant capacity.

The elaboration of new products based on kefir has been carried out. Corona *et al.* (2016) developed new nondairy fermented beverages using vegetable juices as fermentable substrates for kefir water. The study reported that onion, tomato, and strawberry retained their high antioxidant activity after fermentation and that a carrot kefir-like beverage was the product appreciated mainly as a juice. Randazzo and contributors (2016) also reported using fruit juice to make a beverage with kefir sugars in their studies. Apple, quince, grape, kiwifruit, prickly pear, and pomegranate were used, and the results showed the possibility of developing fruit-based kefir-like beverages with high added-value functional properties. WK grains were also used to make a soy whey–based beverage (Tu *et al.* 2019); the novel bioactive soy whey kefir beverage had high functional potential and high acceptability.

7.3 MICROBIAL INTERACTIONS IN KEFIR

Kefir grains are an example of symbiosis between yeast and bacteria. The growth and survival of individual strains are dependent on each other's presence (Farnworth and Farnworth 2005; Gulitz *et al.* 2011; Magalhães-Guedes *et al.* 2010; Miguel *et al.* 2011). Even though both types of MK and WK have different microbial compositions, lactic acid bacteria of the genera *Lactobacillus*, acetic acid bacteria of the genera *Acetobacter*, and *Saccharomyces* yeast appear to be the primary microbial members of kefir grain. Microbial consortia and metabolite production influence the quality and efficacy of the fermented product for human consumption (Gulitz *et al.* 2011; Güzel-Seydim, Gokirmakli, and Greene 2021). Recent studies suggest that microbial community profiling using whole genome shotgun data is feasible, can identify novel species, and can generate a more accurate and detailed assessment of the underlying bacterial community, especially for low-abundance species (Nalbantoglu *et al.* 2014).

During kefir fermentation, more than one type of interaction may co-occur, resulting from the metabolic activity of a succession of different microorganisms. Although in different proportions of the genera of bacteria and yeasts, the symbiotic relationships between MK organisms and WK organisms occur with some similarities (Fiorda *et al.* 2017). The most striking aspect of the symbiosis is the fact that yeasts provide growth factors like amino acids, vitamins, and other compounds for LAB growth, which consequently lead to elevated acid production and thereby optimizes the medium for the growth of the yeast by decreasing the pH (Viljoen 2001; Stadie *et al.* 2013; Van Wyk 2019). On the other hand, the bacterial metabolites of MK, such as lactates, serve as energy sources for the nonlactose fermenting yeasts (i.e., *S. cerevisiae, Y. lipolytica,* and *D. hansenii*), which then raise the pH and stimulate further growth of LAB by preventing potential feedback inhibition by the lactic acid (Rattray and O'Connell 2011). The release of nutrients by yeast cells is an intrinsic characteristic of yeast species and not induced by bacteria (Ponomarova *et al.* 2017; Van Wyk 2019). Ponomarova and contributors (2017) showed that the secretion of amino acids by the yeast is a mechanism to dispose of excess and potentially toxic levels of intracellular nitrogen (i.e., nitrogen overflow). In turn, the secreted amino acids are essential growth-limiting nutrients vital to supporting the growth of the more fastidious LAB. Thus an instantaneous and stable symbiotic relationship is created, which can also be classified as true mutualism.

The interactions between yeasts and LAB also significantly change depending on the sugar source (Leroi and Pidoux 1993). In WK fermentation, sucrose hydrolysis by yeast (i.e., *Saccharomyces, Zygotorulaspora, Dekkera*) via an extracellular β-D-fructofuranosidase (invertase) increases glucose and fructose levels, making this carbon source available to LAB. In addition, part of the ethanol content produced by yeast becomes available for AAB metabolism (Magalhães-Guedes *et al.* 2010). Martínez-Torres *et al.* (2017) proposed insights into metabolic interactions during water kefir fermentation, allowing for the proposal of a minimal and efficient consortium formed by *L. hilgardii, S. cerevisiae,* and *A. tropicalis.* The following metabolic interactions were an initial production of alcohol by *S. cerevisiae,* followed by lactic acid and acetic acid production after 24 h by *Lb. hilgardii* and *A. tropicalis,* respectively; subsequently, acetic acid accumulated due to utilization of ethanol by *A. tropicalis.*

In MK, LAB are primarily responsible for converting the lactose present in milk into lactic acid, which results in a pH decrease and milk preservation. Other kefir microbial constituents include lactose-fermenting yeasts that produce ethanol and CO_2. Nonlactose-fermenting yeast and acetic acid bacteria (AAB) also participate in the process (Magalhães-Guedes *et al.* 2011a). Moreover, ethanol content in MK is low, and usually the alcoholic taste is undetectable. The absence or undetected ethanol might occur due to excessive lactic and acetic acids production by osmophilic lactic acid bacteria, competition for the carbon source, or lysis of the yeast cell walls by bacterial enzymes (Viljoen 2001; Güzel-Seydim *et al.* 2021). Miguel *et al.* (2013) studied MK yeasts and evaluated their potential biotechnological assessment and their ethanol and sugar (glucose/lactose) tolerance.

The kefir flavor is exceptionally complex, and these features are related to the kefir microorganisms, their interactions, and their metabolic products such as lactic and acetic acids, carbon dioxide, ethanol, acetaldehyde, acetoin, and other volatile compounds. Importantly and as has already been discussed, the composition and concentration of the inoculum (grains or cultures), the substrate used, and fermentation conditions should be considered carefully since they significantly influence the dominant strains, diversity, and consequential effects on the beverage characteristics in terms of metabolites produced, concentration, and the flavor and aroma of the final beverage (Lynch *et al.* 2021).

7.4 VOLATILE COMPONENTS

VOCs have been studied because of their possible contribution to kefir's unique taste (Farnworth and Mainville 2003). Several changes can affect the volatile compounds produced in kefir beverage, such as environmental factors, the microbiological composition of kefir grains, the production methods, and the substrate's changing type and fat content. The flavor and aroma of traditional kefir are from the metabolic activity of the bacterial and yeast species present in the kefir grains (Beshkova *et al.* 2003; Cheng 2010; Walsh *et al.* 2016). Generally, more different aroma compounds or fermentation metabolites have been identified in MK than in WK (Tables 7.1 and 7.2). The main two different fermented beverages produced from grains have distinct chemical characteristics. According to the fermentation times shown in Figures 7.3 and 7.4 for MK and WK, changes can be observed, respectively. Acids, ketones, alcohol, aldehydes, esters, and others are the volatile compound groups that predominate in both MK and WK (Tu *et al.* 2019; Wang *et al.* 2021). Although produced in only small amounts (µg/L–mg/L range) in kefir, the volatile organic compounds are essential organoleptically and have been identified as antimicrobial compounds (Gut *et al.* 2021).

TABLE 7.1 VOLATILE COMPOUNDS AND FERMENTATION METABOLITES FOUND IN MILK KEFIR.

Compounds	Substrate	Reference
acetaldehyde, 1-hexanol, 2-methyl-1-butanol, 3-methyl-1-butanol, ethyl acetate	Milk vinegar	Viana *et al.* 2017
2-methyl-1-butanol, 3-methyl-1-butanol, 1-hexanol, 2-methyl-1-propanol, 1-propanol, ethyl acetate, acetaldehyde	Whey	Magalhães *et al.* 2011a
acetaldehyde, diacetyl, acetoin	Cow milk	Güzel-Seydim *et al.* 2000
acetaldehyde, citrate, acetic acid, diacetyl, acetoin, acetate	Cow milk	Grønnevik *et al.* 2011

Compounds	Substrate	Reference
2-propanone, 2-butanone, 2,3-butadione, acetoin, 2-nonanone, 2-undecanoate, ethyl acetate, isoamyl acetate, ethyl hexanoate, ethyl lactate, phenylacetate, ethanol, isoamyl alcohol, phenethyl alcohol, acetic acid, propanoic acid, butanoic acid, pentanoic acid, hexanoic acid, octanoic acid, decanoic acid, benzoic acid, dodecanoic acid	Cow milk	Garofalo *et al.* 2020
isovaleric acid, ethanol, 2-heptanol, 3-methyl-1-butanol, benzene ethanol, 3-methylbutanal, acetaldehyde, carbamic acid, acetic acid, propanoic acid, hexanoic acid, octanoic acid, decanoic acid, 1,2-benzene dicarboxylic acid diethyl ester, 2-hydroxy-isocaproic acid methyl ester, acetic acid ethyl ester, 2-heptanone, 2-butanone, 2-nonanone, 2,6-dimethyl-4-heptanone	Goat and cow milk	Dertli and Çon 2017
acetic acid, 2-methyl propionic acid, 2,3 butanediol, butanoic acid, acetaldehyde, acetone, ethyl acetate, 2-butanone, ethanol, 2,3-butanedione, ethyl butyrate, 2-heptanol, heptanoic acid ethyl ester, 1-propanol 2 -methyl, 3-hydroxy-2-butanone	Cow milk	Aghlara *et al.* 2009
acetaldehyde, acetone, 2-butanone, diacetyl, ethanol, hexanal	Cow milk and soy milk	Liu, Chen, and Lin 2002
acetic acid, hexanoic acid, octanoic acid, nonanoic acid, n-decanoic acid, 2-methyl-1-butanol, 2-ethyl-1-hexanol, ethanol, 2-butanol, 2-methyl-1-propanol, 3-methyl-butanol, phenylethyl alcohol, 3-methyl-butanal, 2-methyl-butanal, octanal, nonanal, pentanal, hexanal, heptanal, ethyl acetate, ethyl butanoate, ethyl hexanoate, 3-methyl-1-butanol acetate, 2,3-butanedione, 2-heptanone, acetone, 2-butanone	Cow milk	Walsh *et al.* 2016

TABLE 7.2 VOLATILE COMPOUNDS AND FERMENTATION METABOLITES FOUND IN WATER KEFIR.

Compounds	Substrate	Reference
acetic acid, hexanoic acid, octanoic acid, decanoic acid, 1-octanol, 2,3-butanediol, 2-ethyl-1-hexanol, phenylethyl alcohol, hexanal, nonanal, 2,4-dimethyl benzaldehyde, acetic acid 2-phenylethyl ester, hexanoic acid ethyl ester, phthalic acid hex3-yl isobutyl ester, decanoic acid, decyl ester, benzene acetic acid ethyl ester	Soy whey	Tu *et al.* 2019
propionic acid, acetic acid, hexanoic acid, decanoic acid, octanoic acid, 2,3-butanediol, 2-ethyl hexanol, 1-hexanol, isoamyl alcohol, phenyl ethyl alcohol, hexanal, nonanal, decanal, 4-methylbenzaldehyde, hexyl acetate, ethyl octanoate, isoamyl acetate, 6-methyl-5-heptene-2-one	Fruit juices	Randazzo *et al.* 2016
acetic acid, decanoic acid, hexanoic acid, octanoic acid, 1-hexanol, phenyl ethyl alcohol, hexanal, nonanal, isoamyl acetate, furfuraldehyde, 5-hydroxymethylfurfural, benzaldehyde, ethyl octanoate, dihydroxyacetone, 1-(3-ethyl phenyl) ethanone	Vegetable juice	Corona *et al.* 2016
2-phenylethyl acetate, ethyl lactate, ethyl butanoate, ethyl 9-decanoate, ethyl benzenepropanoate, methyl octanoate, isoamyl octanoate, diethyl succinate, hexanoate, octanoate, nonanoate, hexanal, furfural, benzaldehyde, 1-octanol, 1,3-propanediol, benzyl alcohol, 2-phenylethanol, 4-ethylphenol, 4-ethyl guaiacol, 2,2-di-tert-butyl phenol, butylated hydroxytoluene, styrene	Sugar and fruit	Laureys and De Vuyst 2017
2-methyl-1-propanol, isoamyl alcohol, ethyl acetate, isoamyl acetate, ethyl hexanoate, ethyl octanoate, ethyl decanoate, manitol	Cane sugar	Laureys and De Vuyst 2014

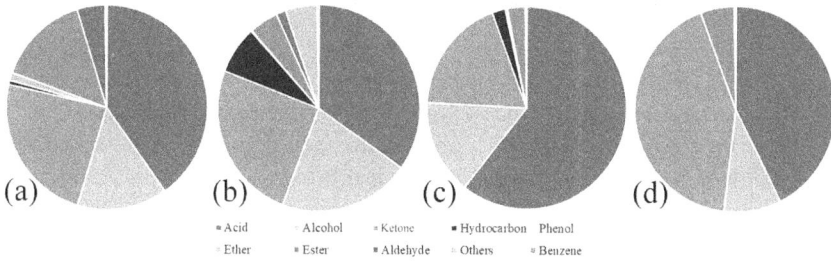

Figure 7.3   Main Volatile Compounds Detected in milk kefir Samples by GC-MS Analysis: (a) Kefir Grains; (b) 12 h of Fermentation; (c) 24 h of Fermentation; (d) 48 h of Kefir Fermentation.

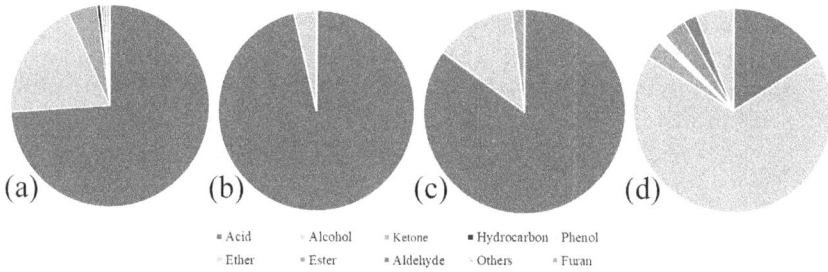

Figure 7.4   Main Volatile Compounds Detected in Water Kefir Samples by GC-MS Analysis: (a) Kefir Grains; (b) 12 h of Fermentation; (c) 24 h of Fermentation; (d) 48 h of Kefir Fermentation.

7.4.1 Volatile Acids

Volatile acidity (VA) describes a group of volatile organic acids of short carbon chain length. Acetic acid usually is identified in WK and MK and constitutes more than 50% of the volatile acids. As a result of fatty acid metabolism by yeast and bacteria, other acids are produced throughout fermentation, such as propionic, hexanoic, nonanoic, and decanoic (Walsh *et al.* 2016; Lynch *et al.* 2021; Garofalo *et al.* 2020).

Acetic acid is of particular importance, and three pathways are responsible for this acid produced in kefir beverage: (1) fermented milk products via heterolactic fermentation by LAB heterofermentative (Randazzo *et al.* 2016); (2) by yeast (i.e., *Saccharomyces*), i.e., the pyruvate dehydrogenase produces acetyl CoA beforehand by the oxidative decarboxylation of pyruvic acid, and the hydrolysis of acetyl CoA can produce acetic acid by acetyl CoA hydrolase

(Ribéreau-Gayon *et al.* 2006); and (3) by the fermentation of carbohydrates in a two-stepped process, which utilizes oxygen to oxidize ethyl alcohol into acetic acid by AAB, which is associated with vinegary flavors (predominantly in WK) (Martínez-Torres *et al.* 2017). Acetic acid was the most representative acid in all the fermented beverages produced by the back slopping method (natural starter cultures using kefir grains), probably due to the high AAB counts found in samples (Garofalo *et al.* 2020).

Acetic, propionic, isobutyric, hexanoic octanoic, decanoic, and hexadecanoic acids were reported in WK-based fruit juices (Randazzo *et al.* 2016). Hexanoic acid is a compound detected in MK and WK and increased in kefir beverage–based juice made from grape, quince, and pomegranate. (Nambou *et al.* 2014) reported that the influential group detected was acid, produced in different concentrations. Hexanoic and octanoic acids were detected in MK-based vegetable juice, and both these organic acids might be associated with the sensory fermented products carrying a refreshing flavor, unique aroma, and texture of the beverage evaluated by Corona *et al.* (2016).

Organic acids may occur in MK and WK due to the hydrolysis of fatty acids, biochemical metabolic process, or bacterial metabolism. Hippuric and uric acids were detected in MK (Güzel-Seydim *et al.* 2000). The authors reported that the hippuric acid disappeared due to the LAB growth in the last 10 h of kefir fermentation. Another organic acid detected in traditionally fermented MK is isovaleric acid. This compound was formed by lipid metabolism and exhibited a sweaty and cheesy odor (Dertli and Çon 2017).

7.4.2 Alcohols

Kefir is a low-alcohol fermented beverage in the MK, while in WK these concentrations may be higher. Different alcohols are naturally produced during microbial fermentation as secondary metabolites (Beshkova *et al.* 2003; Garofalo *et al.* 2020; Bengoa *et al.* 2018). Alcohol production in kefir is mainly due to yeast, and it can be that some *Lactococcus* and *Lactobacillus* species possessing mild alcohol dehydrogenase activity produce these compounds (Magalhães *et al.* 2011a; Walsh *et al.* 2016). Higher alcohols can positively and negatively impact the aroma and flavor in MK and WK beverages. For example, 2-heptanol is responsible for the buttery flavor and fruity odor (Garofalo *et al.* 2020; Guangsen *et al.* 2021). Phenethyl alcohol confers a rose, violet-like, honey, floral, spicy flavor (Randazzo *et al.* 2016; Mota-Gutierrez *et al.*, 2019). 2,3-Butanediol produces a buttery, creamy, and fruity flavor (Beshkova *et al.* 2003), and it is produced by LAB via the butanediol fermentation pathway. Isoamyl alcohol is described as a harsh nail polish remover (Laureys and De Vuyst 2014) and was detected in a cheese whey–based beverage using kefir grains (Magalhães-Guedes *et al.* 2011a). Phenethyl alcohol is synthesized by yeast, mainly species of the genus *Kluyveromyces*, *Saccharomyces*, and *Pichia*, but it can also be produced by *Lactobacillus* species (Walsh *et al.* 2016). These compounds can influence the flavor and aroma of the Kefir beverage, but they also have an antimicrobial function (Gut *et al.* 2021). The alcohols 1-hexanol and 3-methyl-1-butanol have a positive influence (20 mg/L) on the aroma of the MK and WK beverage; on the other hand, >20 mg/L concentrations of these alcohols, having a volatile description of "coconut-like," "harsh," and "pungent," can contribute negatively to the product aroma (Dragone *et al.* 2009; Gómez-Míguez *et al.* 2007). Another volatile higher alcohol identified in MK and whey kefir is 2-methyl-1-butanol (Magalhães-Guedes *et al.* 2011a). This compound is produced during the catabolism of the branched-chain amino acid isoleucine. A higher concentration of 2-methyl-1-butanol in the whey-based beverages could be related to the higher isoleucine content.

The production of ethanol in kefir is complex, and both microorganism yeasts and heterofermentative bacteria can produce it (Beshkova *et al.* 2003). *Lactococcus* and *Lactobacillus* species possess mild alcohol dehydrogenase activity that converts acetaldehyde to ethanol. The quantity of ethanol produced is dependent on the fermentation process and the type of container used (tightly capped or not). Ethanol is a metabolite of the alcohol fermentation produced by yeasts during the fermenting of MK and WK beverages (Garofalo *et al.* 2020). The growth of AAB benefits from increased ethanol production to acetic acid metabolism (Fiorda *et al.* 2017; Norberto *et al.* 2018). The ethanol concentration can be 0.5–1.5%. The final alcohol concentration is determined for the most part by the number of yeasts present in the grains added to the milk and the time of fermentation. Magalhães-Guedes and contributors (2011c) reported that the alcohol content should be enough to give kefir the flavor of a light alcoholic beverage typical of traditional (ancient) kefir of the Caucasus; the yeast aroma ensures the specificity of this type of fermented beverage. Another factor influencing this concentration may vary with whether MK or WK is used and the production method used because fermentations are interrupted at higher pH levels, generating a lower ethanol concentration (Farnworth and Mainville 2003).

7.4.3 Aldehydes

Aldehydes are chemically a compound containing a functional group with the structure –CHO, consisting of a carbonyl center (a carbon double-bonded to oxygen) with the carbon atom also bonded to hydrogen generic alkyl or side chain R group (Belitz *et al.* 2009). Aldehydes and ketones are usually found in fermented beverages as by-products of yeast fermentation, intermediates in fuel oil formation, and alcohol oxidation at various stages of beverage production (Magalhães-Guedes *et al.* 2011a). In kefir fermentation, aldehydes formed from lipid oxidation are subjected to further modification by microbial reductases or amino acids enzymolysis (Aghlara *et al.* 2009; Walsh *et al.* 2016; Wang *et al.* 2021). Walsh and contributors (2016) reported that only two aldehydes (hexanal and benzaldehyde) were detected used goat MK and in low concentrations indicated an optimal condition because they may play off the flavor (aldehydic and bitter) of kefir. Hexanal is an aldehyde commonly found in MK and WK beverages and is associated with a different flavor. In soybean kefir, it presented a grassy flavor (Tu *et al.* 2019), but it can present a green, slightly fruity, lemon, herbal, grassy, tallow flavor in MK (Walsh *et al.* 2016). Nonanal was the main aldehyde detected in WK-based fruit juice, followed by benzaldehyde (Randazzo *et al.* 2016). Nonanal is produced through lipid metabolism by *S. cerevisiae* and is associated with green, citrus, fatty, and floral odors (Walsh *et al.* 2016). Benzaldehyde was also reported in bread produced by kefir sourdough, revealing a positive contribution (Plessas *et al.* 2011). Other aldehydes that can also be formed during the kefir fermentation are octanal, pentanal, hexanal, and heptanal and are produced by lipid metabolism, producing an odor of relevant green, slightly fruity, lemon, herbal, grassy, tallow, fatty, apple, malty woody (Walsh *et al.* 2016). Aldehydes 3-methyl-butanal and 2-methyl-butanal are examples produced by amino acid metabolism and give malty, cheesy, green, dark chocolate, cocoa flavors (Walsh *et al.* 2016; Dertli and Çon 2017). Hexanal is an aldehyde detected in milk and soymilk kefir during 24 h of fermentation (Liu *et al.* 2002). This compound is produced from the breakdown of polyunsaturated fatty acids catalyzed by a lipoxygenase and is deemed responsible for the unpleasant spoiled flavor of soymilk.

Acetaldehyde is one of the most critical contributors to flavor in kefir beverage and was suggested to be a representative aldehyde compound in MK and WK (Beshkova *et al.* 2003; Aghlara *et al.* 2009; Dertli and Çon 2017). Acetaldehyde can be detected in low concentration, and this is explained by the ability of this compound to be converted to ethanol by alcohol dehydrogenase, besides being formed either by the lactate metabolism or by ethanol oxidation (Nambou *et al.* 2014). This compound was reported in low concentration in MK made with kefir grains (Güzel-Seydim *et al.* 2000) and produced by using pure kefir starter cultures, mainly *Lactobacillus delbrueckii* (Nambou *et al.* 2014). Acetaldehyde accumulation in the growth medium can occur when the specific activities of enzymes that form acetaldehyde are greater than the alcohol dehydrogenase's ability to convert it to ethanol (Aghlara *et al.* 2009). The main stage of acetaldehyde formation occurred during the first hours of the fermentation of MK and WK. During this time, LAB from the starter culture metabolizes lactose to lactate (Beshkova *et al.* 2003). Furthermore, acetaldehyde is a compound significantly associated with determining the flavor-aroma characteristic of yogurt-type products and has also been detected in cheese whey–based beverages using kefir grains (Magalhães-Guedes *et al.* 2011a, 2011b).

7.4.4 Esters

Esters are a group of volatile compounds that can significantly affect the fruity and floral flavors; they are formed when an alcohol function reacts with an acid function, and a water molecule is eliminated (Ribéreau-Gayon *et al.* 2006). The esters found in kefir can be produced both by LAB and yeasts during fermentation as secondary products of sugar metabolism and constitute one of the most significant and most essential compounds affecting flavor (Beshkova *et al.* 2003; Walsh *et al.* 2016). The most significant esters found in MK and WK are ethyl acetate (fruity, pineapple, sweet), ethyl butanoate (fruity, banana, sweet), ethyl hexanoate (fruity, apple, banana), ethyl octanoate (fruity, banana, pineapple), ethyl decanoate (fruity, grape, cognac), isoamyl acetate (fruity, banana), phenylacetate (fruity, sweet, honey) (Laureys and De Vuyst 2014; Randazzo *et al.* 2016; Walsh *et al.* 2016; Garofalo *et al.* 2020; Güzel-Seydim *et al.* 2021).

All alcohols and acids may react to form esters (Peddie 1990; Swiegers *et al.* 2005). Alcoholysis is essentially a transferase reaction in which fatty acyl groups from acylglycerols and acyl CoA derivatives are directly transferred to alcohols and are the primary mechanism of ester biosynthesis by lactic acid bacteria and yeasts (Cristiani and Monnet 2001; Liu, Holland, and Crow 2004). Three key factors determine ester biosynthesis by dairy microorganisms: enzymes, substrates, and environment (Liu, Holland, and Crow 2004). Due to the symbiotic relationships and microbial composition in kefir, the biosynthetic rates of esters should be affected not only by the presence of microorganisms but also by concentrations of metabolic precursors and the enzymes involved in synthesis and degradation (Hu *et al.* 2014). In a previous study, *L. kefiranofaciens*, *Lc. lactis*, and *S. cerevisiae* were demonstrated to have strong correlations with esters production in kefir, mainly ethyl acetate (fruity flavors) (Beshkova *et al.* 2003; Walsh *et al.* 2016; Garofalo *et al.* 2020).

Ethyl acetate has a significant effect on the organoleptic characteristics of fermented beverages (Liu, Holland, and Crow 2004; Swiegers *et al.* 2005). It can contribute positively at low concentrations, resulting in a pleasant aroma with fruity properties, but it can turn vinegary at levels above 150 mg/L, adding spoilage notes to the beverage (Falqué *et al.* 2001). Ethyl acetate concentration in kefir beverage performed by inoculating grains in milk (ML), cheese whey (CW), and deproteinized cheese whey (DCW) were below 10 mg/L, a level suitable to confer a pleasant flavor (Magalhães-Guedes *et al.* 2011a, 2011b) Ethyl acetate was not detected in kefir beverages by other authors (Güzel-Seydim *et al.* 2000; Randazzo *et al.* 2016; Tu *et al.* 2019).

7.4.5 Ketones

Several molecules with ketone functions have been identified in MK and WK as by-products of yeast and bacteria fermentation, including 2,3-butanedione,

2,3-pentanedione, 3-hydroxy-2-butanone, 2-butanone, acetone, 2-heptanone, 2-nonanone, among others, depending on the substrate, composition of grains or cultures, and the production methods (Beshkova *et al.* 2003; Aghlara *et al.* 2009; Walsh *et al.* 2016; Garofalo *et al.* 2020). Collectively known as C4 compounds, these include 2,3-butanedione (diacetyl), 3-hydroxy-2-butanone (acetoin), and 2- butanone (Mayo *et al.* 2016). They can be generated from glycolysis or citrate metabolism of microorganisms (Swiegers *et al.* 2005) and are responsible for the typical aroma of yogurt with its butter-like flavor (Cheng 2010), these could modulate the sensory characteristics of the MK and WK.

The main flavor compound is 2,3-butanedione, commonly referred to as diacetyl, the ketone with the pronounced trademark buttery aroma (Swiegers *et al.* 2005). Small quantities of diacetyl, ranging from 0.3 to 1.0 mg/L, contribute to kefir's pleasant and delicate flavor and aroma (Van Wyk 2019). The diacetyl formation and degradation are causally related to the LAB growth and glucose catabolism, and its production relates to pyruvate formation. The metabolism of pyruvate can yield in LAB different end products such as lactate, formate, acetate, ethanol, and the aroma compounds of four carbons (C4 compounds) diacetyl, acetoin, and butanediol (Mayo *et al.* 2016). Two bacterial mechanisms have been suggested: (1) decarboxylation of α-acetolactate, which is formed by the α-acetolactate synthase (α-ALS), direct synthesis from acetyl coenzyme-A and aceylthiamin pyrophosphate (acetyl-TPP) or (2) the condensation reaction of acetaldehyde or pyruvate, catalyzed by pyruvate dehydrogenase complex (PDC) with hydroxyethyl-TPP as an intermediate product (Snoep *et al.* 1992; von Wright and Axelsson 2019). In milk, the pyruvate surplus is provided by the breakdown of citrate, typically present in significant amounts (von Wright and Axelsson 2019). Diacetyl was reported in Kefir (Beshkova *et al.* 2003; Aghlara *et al.* 2009; Garofalo *et al.* 2020). In a previous study, Walsh *et al.* 2016) reported the correlation of diacetyl production with *Leuconostoc mesenteroides* and *Lactobacillus helveticus*. This compound was not detected in kefir by other authors (Güzel-Seydim *et al.* 2000; Magalhães-Guedes *et al.* 2011a, 2011b; Corona *et al.* 2016; Randazzo *et al.* 2016; Dertli and Çon 2017; Tu *et al.* 2019; Chen *et al.* 2021; Wang *et al.* 2021).

Acetoin, or 3-hydroxy-2-butanone, is a common flavor substance in many fermented foods. Acetoin has a mild creamy, slightly sweet, butter-like flavor that is like that of diacetyl. The flavor of acetoin is considerably weaker than that of diacetyl and can impart a mild, pleasant, buttery taste (Cheng 2010). Acetoin can also readily convert from diacetyl by the enzyme diacetyl reductase (DAR). This enzyme also possesses acetoin reductase activity, yielding 2,3-butanediol from acetoin, while the reverse reaction is catalyzed by 2,3-butanediol dehydrogenase (BDH) (Mayo *et al.* 2016). Typical acetoin concentrations in kefir can range, as Aghlara *et al.* (2009) observed, to the highest concentration in MK of 67.8 mg/L, followed by a reduction to a final concentration of 53.1 mg/L. Randazzo *et al.* (2016) used fruit juices to ferment with WK and found only around 9 µg/L of acetoin in kefir prepared with quince. On the other hand, acetoin was not detected in kefir by other authors (Beshkova *et al.* 2003; Magalhães-Guedes *et al.* 2011a; Walsh *et al.* 2016; Tu *et al.* 2019; Chen *et al.* 2021; Wang *et al.* 2021).

Acetone and 2-butanone are described as volatile compounds of less importance when compared to diacetyl and acetoin. Both are responsible for influencing the characteristic flavor of fermented milk. Acetone has a sweet, fruity aroma, and 2-butanone is a characteristic flavor component of fermented milk (Cheng 2010). In MK, the presence of 2-butanone was related to the production ability of *Lb. helveticus*, and the presence of acetone in kefir was due to the *Lb. bulgaricus* and *Lb. helveticus* (Beshkova *et al.* 2003).

7.5 FINAL CONSIDERATIONS AND FUTURE PERSPECTIVES

Kefir has been gaining high levels of popularity over the last few years, owing to its nutritional benefits and its unique taste. Volatile compounds such as acids, ketones, alcohol, aldehydes, and esters in MK and WK result from the metabolic activity of several bacterial and yeast species that make up the grains and contribute to kefir's distinctive flavor.

Although the industrial production of the beverage is not yet high, the continuous innovation and preservation of sensory characteristics are still a challenge to drive the growth of the kefir products market.

REFERENCES

Aghlara, A., S. Mustafa, Y. A. Manap, and R. Mohamad. 2009. Characterization of headspace volatile flavor compounds formed during kefir production: Application of solid phase microextraction. *International Journal of Food Properties* 12:808–818.

Altay, F., F. Karbancioglu-Güler, C. Daskaya-Dikmen, and D. Heperkan. 2013. A review on traditional turkish fermented non-alcoholic beverages: Microbiota, fermentation process and quality characteristics. *International Journal of Food Microbiology* 167:44–56.

Angulo, L., E. Lope and C. Lema. 1993. Microflora present in kefir grains of the galician region North-West of Spain. *Journal of Dairy Research* 60:263–267.

Arslan, S. 2015. A review: Chemical, microbiological and nutritional characteristics of kefir. *CYTA—Journal of Food* 13:340–345.

Barbosa, A. F., P. G. Santos, A. M. S. Lucho, and J. M. Schneedorf. 2011. Kefiran can disrupt the cell membrane through induced pore formation. *Journal of Electroanalytical Chemistry* 653:61–66.

Belitz, H. D., W. Grosch, and P. Schieberle. 2009. *Food Chemistry: Amino Acids, Peptides, Proteins*. Berlin and Heidelberg: Springer.

Bengoa, A. A., L. Zavala, P. Carasi *et al.* 2018. Simulated gastrointestinal conditions increase adhesion ability of *Lactobacillus Paracasei* strains isolated from kefir to caco-2 cells and mucin. *Food Research International* 103:462–467.

Beshkova, D. M., E. D. Simova, G. I. Frengova, Z. I. Simov, and Z. P. Dimitrov. 2003. Production of volatile aroma compounds by kefir starter cultures. *International Dairy Journal* 13:529–535.

Bourrie, B. C. T., B. P. Willing, and P. D. Cotter. 2016. The microbiota and health promoting characteristics of the fermented beverage kefir. *Frontiers in Microbiology* 7:647.

Cais-Sokolińska, D., J. Wójtowski, J. Pikul *et al.* 2015. Formation of volatile compounds in kefir made of goat and sheep milk with high polyunsaturated fatty acid content. *Journal of Dairy Science* 98:6692–6705.

Chen, Z., T. Liu, T. Ye *et al.* 2021. Effect of lactic acid bacteria and yeasts on the structure and fermentation properties of Tibetan kefir grains. *International Dairy Journal* 114:104943.

Cheng, H. 2010. Volatile flavor compounds in yogurt: A review. *Critical Reviews in Food Science and Nutrition* 50:938–950.

Chunchom, S., C. Talubmook, and S. Deeseenthum. 2017. Antioxidant activity, biochemical components and sub-chronic toxicity of different brown rice kefir powders. *Pharmacognosy Journal* 9:388–394.

Corona, O., W. Randazzo, A. Miceli *et al.* 2016. Characterization of kefir-like beverages produced from vegetable juices. *LWT—Food Science and Technology* 66:572–581.

Cristiani, G., and V. Monnet. 2001. Food micro-organisms and aromatic ester synthesis. *Sciences Des Aliments* 21:211–230.

Cui, X. H., S. J. Chen, Y. Wang, Y. and J. R. Han. 2013. Fermentation conditions of walnut milk beverage inoculated with kefir grains. *LWT—Food Science and Technology* 50:349–352.

Dertli, E., and A. H. Çon. 2017. Microbial diversity of traditional kefir grains and their role on kefir aroma. *LWT—Food Science and Technology* 85:151–157.

Dias, P. A., D. T. Silva, T. S. Tejada, M. C. G. M. Leal, R. C. S. Conceição, and C. D. Timm. 2012. Survival of pathogenic microorganisms in kefir. *Revista do Instituto Adolfo Lutz* 71:182–186.

Dragone, G., S. I. Mussatto, J. M. Oliveira, and J. A. Teixeira. 2009. Characterisation of volatile compounds in an alcoholic beverage produced by whey fermentation. *Food Chemistry* 112:929–935.

Falqué, E., E. Fernández, and D. Dubourdieu. 2001. Differentiation of white wines by their aromatic index. *Talanta* 54:271–281.

Farnworth, E. R., and E. R. Farnworth. 2005. Kefir—a complex probiotic. Edited by G. R. Gilbson. *Food Science and Technology Bulletin: Functional Foods* 21: 1–17. International Food Information Service.

Farnworth, E. R., and I. Mainville. 2003. Kefir—a fermented milk product. Edited by E. R. Farnworth. *Handbook of Fermented Functional Foods*, 2nd ed., 89–127. CRC Press

Fiorda, F. A., G. V. M. Pereira, V. T. Soccol *et al.* 2017. Microbiological, biochemical, and functional aspects of sugary kefir fermentation—a review. *Food Microbiology* 66:86–95.

Garofalo, C., I. Ferrocino, A. Reale *et al.* 2020. Study of kefir drinks produced by backslopping method using kefir grains from bosnia and herzegovina: Microbial dynamics and volatilome profile. *Food Research International* 137:109369.

Garrote, G. L., A. G. Abraham, and G. L. Antoni. 2001. Chemical and microbiological characterisation of kefir grains. *Journal of Dairy Research* 68:639–652.

Gaware, V. M., K. Kotade, R. Dolas *et al.* 2011. The magic of kefir: A review. *Pharmacology online* 1:376–386.

Gómez-Míguez, M. J., J. F. Cacho, V. Ferreira, I. M. Vicario, and F. J. Heredia. 2007. Volatile components of zalema white wines. *Food Chemistry* 100:1464–1473.

Grønnevik, H., M. Falstad, and J. Narvhus. 2011. Microbiological and chemical properties of Norwegian kefir during storage. *International Dairy Journal* 21:601–606.

Guangsen, T., L. Xiang, and G. Jiahu. 2021. Microbial diversity and volatile metabolites of kefir prepared by different milk types. *CyTA—Journal of Food* 19:399–407.

Gulitz, A., J. Stadie, M. Wenning, M. A. Ehrmann and R. F. Vogel. 2011. The microbial diversity of water kefir. *International Journal of Food Microbiology* 151:284–288.

Gut, A. M., T. Vasiljevic, T. Yeager, and O. N. Donkor. 2021. Kefir characteristics and antibacterial properties—potential applications in control of enteric bacterial infection. *International Dairy Journal* 118:105021–105032.

Güzel-Seydim, Z. B., Ç. Gokirmakli, and A. K. Greene. 2021. A comparison of milk kefir and water kefir: Physical, chemical, microbiological and functional properties. *Trends in Food Science & Technology* 113:42–53.

Güzel-Seydim, Z. B., A. C. Seydim, A. K Greene, and A. B. Bodine. 2000. Determination of organic acids and volatile flavor substances in kefir during fermentation. *Journal of Food Composition and Analysis* 13:35–43.

Hsieh, H. H., S. Y. Wang, T. L Chen, Y. L Huang, and M. J Che. 2012. Effects of cow's and goat's milk as fermentation media on the microbial ecology of sugary kefir grains. *International Journal of Food Microbiology* 157:73–81.

Hu, J. B., S. Gunathilake, Y. C. Chen, and P. L. Urban. 2014. On the dynamics of kefir volatome. *Royal Society of Chemistry* 4:28865–28870.

Huang, Y., F. Wu, X. Wang, Y. Sui, L. Yang, and J. Wang. 2013. Characterization of lactobacillus plantarum Lp27 isolated from tibetan kefir grains: A potential probiotic bacterium with cholesterol-lowering effects. *Journal of Dairy Science* 96:2816–2825.

Kim, D. H., D. Jeong, H. Kim *et al.* 2016. Antimicrobial activity of kefir against various food pathogens and spoilage bacteria. *Korean Journal for Food Science of Animal Resources* 36:787–790.

Laureys, D., and L. De Vuyst. 2014. Microbial species diversity, community dynamics, and metabolite kinetics of water kefir fermentation. *Applied and Environmental Microbiology* 80:2564–272.

Laureys, D., and L. De Vuyst. 2017. The water kefir grain inoculum determines the characteristics of the resulting water kefir fermentation process. *Journal of Applied Microbiology* 122:719–732.

Lee, M. Y., K. S. Ahn, O. K. Kwon *et al.* 2007. Anti-inflammatory and anti-allergic effects of kefir in a mouse asthma model. *Immunobiology* 212:647–654.

Leroi, F., and P. Pidoux. 1993. Detection of interactions between yeasts and lactic acid bacteria isolated from sugary kefir grains. *Journal of Applied Bacteriology* 74:48–53.

Liu, J. R., M. J. Chen, and C. W. Lin. 2002. Characterization of polysaccharide and volatile compounds produced by kefir grains grown in soymilk. *Journal of Food Science* 67:104–108.

Liu, J. R., S. Y. Wang, Y. Y. Lin, and C. W. Lin. 2002. Antitumor activity of milk kefir and soy milk kefir in tumor-bearing mice. *Nutrition and Cancer* 44:183–187.

Liu, S. Q., R. Holland, and V. L. Crow. 2004. Esters and their biosynthesis in fermented dairy products: A review. *International Dairy Journal* 14:923–945.

Lynch, K. M., S. Wilkinson, L. Daenen, and E. K. Arendt. 2021. An update on water kefir: Microbiology, composition and production. *International Journal of Food Microbiology* 345:109128.

Magalhães-Guedes, K. T., I. T. Barreto, P. P. L. G. Tavares *et al.* 2020. Effect of kefir biomass on nutritional, microbiological, and sensory properties of mango-based popsicles. *International Food Research Journal* 27:536–545.

Magalhães-Guedes, K. T., D. R. Dias, G. V. M. Pereira *et al.* 2011a. Chemical composition and sensory analysis of cheese whey-based beverages using kefir grains as starter culture. *International Journal of Food Science and Technology* 46:871–878.

Magalhães-Guedes, K. T., G. Dragone, G. V. M. Pereira *et al.* 2011b. Comparative study of the biochemical changes and volatile compound formations during the production of novel whey-based kefir beverages and traditional milk kefir. *Food Chemistry* 126:249–253.

Magalhães-Guedes, K. T., G. V. M. Pereira, C. R. Campos, G. Dragone, and R. F. Schwan. 2011c. Brazilian kefir: Structure, microbial communities and chemical composition. *Brazilian Journal of Microbiology* 42:693–702.

Magalhães-Guedes, K. T., G. V. M. Pereira, D. R. Dias, and R. F. Schwan. 2010. Microbial communities and chemical changes during fermentation of sugary brazilian kefir. *World Journal of Microbiology and Biotechnology* 26:1241–1250.

Marsh, A. J., O. O'Sullivan, C. Hill, R. P. Ross, and P. D. Cotter. 2013. Sequencing-based analysis of the bacterial and fungal composition of kefir grains and milks from multiple sources. *PLoS One* 8:e69371.

Marshall, V. M. and W. M. Cole. 1985. Methods for making kefir and fermented milks based on kefir. *Journal of Dairy Research* 52:451–456.

Martínez-Torres, A., S. Gutiérrez-Ambrocio, P. Heredia-del-Orbe, L. Villa-Tanaca, and C. Hernández-Rodríguez. 2017. Inferring the role of microorganisms in water kefir fermentations. *International Journal of Food Science and Technology* 52:559–571.

Mayo, B., T. Aleksandrzak-Piekarczyk, M. F. M. Kowalczyk *et al.* 2016. Updates in the metabolism of lactic acid bacteria. In *Biotechnology of lactic acid bacteria—novel applications*, edited by F. Mozzi, R. R. Raya, and G. M. Vignolo, 3–33. Chichester: Wiley-Blackwell Publishing.

Meydani, S. N. and W. K. Ha. 2000. Immunologic effects of yogurt. *American Journal of Clinical Nutrition* 71:861–872.

Miguel, M. G. C. P., P. G. Cardoso, L. A. Lago, and R. F. Schwan. 2010. Diversity of bacteria present in milk kefir grains using culture-dependent and culture-independent methods. *Food Research International* 43:1523–1528.

Miguel, M. G. C. P., P. G. Cradoso, K. T. Magalhães-Guedes, and R. F. Schwan. 2011. Profile of microbial communities present in Tibico sugary kefir grains from different brazilian states. *World Journal of Microbiology and Biotechnology* 27:1875–1884.

Miguel, M. G. C. P., P. G. Cardoso, K. T. Magalhães-Guedes, and R. F. Schwan. 2013. Identification and assessment of kefir yeast potential for sugar/ethanol-resistance. *Brazilian Journal of Microbiology* 44:113–118.

Mota-Gutierrez, J., L. Barbosa-Pereira, I. Ferrocino, and L. Cocolin. 2019. Traceability of functional volatile compounds generated on inoculated cocoa fermentation and its potential health benefits. *Nutrients* 11:884.

Nalbantoglu, U., A. Cakar, H. Dogan *et al.* 2014. Metagenomic analysis of the microbial community in kefir grains. *Food Microbiology* 4:42–51.

Nambou, K., C. Gao, F. Zhou, B. Guo, L. Ai, and Z. J. Wu. 2014. A novel approach of direct formulation of defined starter cultures for different kefir-like beverage production. *International Dairy Journal* 34:237–246.

Norberto, A. P., R. P. Marmentini, P. H. Carvalho *et al.* 2018. Impact of partial and total replacement of milk by water-soluble soybean extract on fermentation and growth parameters of kefir microorganisms. *LWT* 93:491–498.

Peddie, H. A. B. 1990. Ester formation in brewery fermentations. *Journal of the Institute of Brewing* 96: 327–331.

Plessas, S., A. Alexopoulos, C. Voidarou, E. Stavropoulou, and E. Bezirtzoglou. 2011. Microbial ecology and quality assurance in food fermentation systems: The case of kefir grains application. *Anaerobe* 17:483–485.

Plessas, S., C. Nouska, I. Mantzourani, Y. Kourkoutas, A. Alexopoulos, and E. Bezirtzoglou. 2017. Microbiological exploration of different types of kefir grains. *Fermentation* 3:1.

Plessas, S., L. Pherson, A. Bekatorou, P. Nigam, and A. A. Koutinas. 2005. Bread making using kefir grains as baker's yeast. *Food Chemistry* 93:585–589.

Ponomarova, O., N. Gabrielli, D. C. Sévin *et al.* 2017. Yeast creates a niche for symbiotic lactic acid bacteria through nitrogen overflow. *Cell Systems* 5:345–357.

Puerari, C., K. T. Magalhães, and R. F. Schwan. 2012. New cocoa pulp-based kefir beverages: Microbiological, chemical composition and sensory analysis. *Food Research International* 48:634–640.

Randazzo, W., O. Corona, R. Guarcello *et al.* 2016. Development of new non-dairy beverages from mediterranean fruit juices fermented with water kefir microorganisms. *Food Microbiology* 54:40–51.

Rattray, F. P., and M. J. O'Connell. 2011. Fermented Milks|Kefir. In *Encyclopedia of Dairy Sciences,* 2nd ed., 518–524. Academic Press.

Ribéreau-Gayon, P., Y. Glories, A. Maujean, and D. Dubourdieu. 2006. *Handbook of Enology, the Chemistry of Wine: Stabilization and Treatments*, 2nd ed., vol. 2. Chichester: Wiley.

Rodrigues, K. L., J. C. T. Carvalho, and J. M. Schneedorf. 2005. Anti-inflammatory properties of kefir and its polysaccharide extract. *Inflammopharmacology* 13:485–492.

Schwan, R. F., K. T. Magalhães-Guedes, and D. R. Dias. 2015. Kefir—grains and beverages: A review. *Scientia Agraria Paranaensis* 14:1–9.

Simova, E., D. Beshkova, A. Angelov, T. Hristozova, G. Frengova, and Z. Spasov. 2002. Lactic acid bacteria and yeasts in kefir grains and kefir made from them. *Journal of Industrial Microbiology and Biotechnology* 28:1–6.

Snoep, J. L., M. J. Teixeira de Mattos, M. J. Starrenburg, and J. Hugenholtz. 1992. Isolation, characterization, and physiological role of the pyruvate dehydrogenase complex and ac-acetolactate synthase of *Lactococcus lactis* subsp. lactis bv. *diacetylactis. Journal of Bacteriology* 174:4838–4841.

Souza, B. L., K. T. Magalhães-Guedes, P. V. F. Lemos *et al.* 2020. Development of arrowroot flour fermented by kefir grains. *Journal of Food Science* 85:3722–3730.

Stadie, J., A. Gulitz, M. A. Ehrmann, and R. F. Vogel. 2013. Metabolic activity and symbiotic interactions of lactic acid bacteria and yeasts isolated from water kefir. *Food Microbiology* 35:92–98.

Swiegers, J. H., E. J. Bartowsky, P. A. Henschke, and I. S. Pretorius. 2005. Yeast and bacterial modulation of wine aroma and flavour. *Australian Journal of Grape and Wine Research* 11:139–1373.

Takizawa, S., S. Kojima, S. Tamura, S. Fujinaga, Y. Benno, and T. Nakase. 1998. The composition of the *Lactobacillus* flora in kefir grains. *Systematic and Applied Microbiology* 21:121–127.

Tu, C., F. Azi, J. Huang, X. Xu, G. Xing, and M. Dong. 2019. Quality and metagenomic evaluation of a novel functional beverage produced from soy whey using water kefir grains. *LWT—Food Science and Technology* 113:108258–108268.

Van Wyk, J. 2019. Kefir: The champagne of fermented beverages. *Fermented Beverages: Volume 5. The Science of Beverages:* 473–527.

Viana, R. O., K. T. Magalhães-Guedes, R. A. Braga, D. R. Dias, and R. F. Schwan. 2017. Fermentation process for production of apple-based kefir vinegar: Microbiological, chemical and sensory analysis. *Brazilian Journal of Microbiology* 48:592–601.

Viljoen, B. C. 2001. The interaction between yeasts and bacteria in dairy environments. In *International Journal of Food Microbiology* 69:37–44.

von Wright, A., and L. Axelsson. 2019. Lactic acid bacteria: An introduction. In *Lactic Acid Bacteria: Microbiological and Functional Aspects*, edited by Gabriel Vinderola, 1–16. New York: CRC Press, Taylor & Francis Group.

Walsh, A. M., F. Crispie, K. Kilcawley *et al.* 2016. Microbial succession and flavor production in the fermented dairy beverage kefir. *MSystems* 1:1–17.

Wang, H., X. Sun, X. Song, and M. Guo. 2021. Effects of kefir grains from different origins on proteolysis and volatile profile of goat milk kefir. *Food Chemistry* 339:128099–128108.

Witthuhn, R. C., and T. Schoeman. 2004. Isolation and characterization of the microbial population of different South African kefir grains. *International Dairy Journal* 57:33–37.

Witthuhn, R. C., T. Schoeman, and T. Britz. 2005. Characterisation of the microbial population at different stages of kefir production and kefir grain mass cultivation. *International Dairy Journal* 15:383–389.

Zhang, J., X. Zhao, Y. Jiang *et al.* 2017. Antioxidant status and gut microbiota change in an aging mouse model as influenced by exopolysaccharide produced by *Lactobacillus Plantarum* YW11 isolated from tibetan kefir. *Journal of Dairy Science* 100:6025–6041.

Chapter 8

Volatile Compounds Formation in Kombucha

Jasmina Vitas, Radomir Malbaša, and Stefan Vukmanović

CONTENTS

ACKNOWLEDGMENT

The authors want to thank to the Ministry of Education, Science and Technological Development of Republic of Serbia, Grant Number: 451–03–9/2021–14/200134, for financial support.

8.1 INTRODUCTION

Kombucha is the symbiotic association between yeasts and bacteria. It is capable of converting a simple substrate, usually black or green tea sweetened with sucrose, into a sweet, slightly carbonated, and refreshing beverage. Kombucha beverage, besides its refreshing properties, could be used as an auxiliary remedy. The most common way to prepare the kombucha beverage is by domestic production. Sweet, slightly carbonated and refreshing kombucha beverage is consumed in several countries of the world, but traditionally the main consumers have been in China, Russia, and Germany (Greenwalt, Steinkraus, and Ledford 2000; Jayabalan *et al.* 2014; Zhang 2019).

The Latin name of the kombucha culture is *Medusomyces gisevii*, and the composition of microbes in the kombucha culture is not the same around the world (Jayabalan, Malbaša, and Sathishkumar 2015; Blanc 1996; Lindau 1913).

The chemical composition of kombucha beverage is very versatile, being composed of sugars, acids, vitamins, enzymes, essential ions, alcohol, and other components (Malbaša 2004). The amount and the presence of volatile compounds determine the odor and taste of a given food product and therefore impact the overall quality. Although there are numerous volatiles, it was established that only a small number of them actually influence odor (Maarse 1991).

This chapter aims to provide a literature overview of the formation of the volatile compounds in different types of kombucha beverages.

8.2 KOMBUCHA HISTORY

The first facts about kombucha consumption are dated to 220 BCE during the rule of the Tsin dynasty in Manchuria. The trade routes spread kombucha to Russia and Eastern Europe. Kombucha became popular in Russia, where it was used to treat different metabolic diseases, hemorrhoids, and rheumatism. A lower rate of carcinogenic illness has been recorded in Russian regions where kombucha beverage was regularly consumed after World War II despite the high pollution caused by war (Frank 1995; Jayabalan *et al.* 2014).

During World War II, kombucha was transferred from Russia to Western Europe and Northern Africa. Europeans mostly used kombucha because of its detoxifying properties (Blanc 1996).

Throughout history, it was known by different names like kombucha, tea fungus, Indian tea fungus, Manchurian fungus, hongo, Russian jellyfish, haipao, *champignon de longue vie*, among others (Frank 1995). The first system name for kombucha was *Medusomyces gisevii* (Lindau 1913), and the second system name, *Auricularia delicata*, was mentioned by Filho *et al.* (1985). Both system names are wrong because *Medusomyces gisevii* is related just to yeasts, and genus *Auricularia* covers mushrooms. However, kombucha is the most accepted name, and people commonly use it.

8.3 KOMBUCHA PRODUCTION

The common substrate for kombucha cultivation is black or green tea water extract sweetened with sucrose, and the obtained products are known as traditional kombucha products. The usual amount of sucrose in black tea substrate is 5–8% (Reiss 1987), and the usual tea content is 1.5–4.5 g/L (Sievers *et al.* 1995; Greenwalt, Ledford, and Steinkraus 1998; Malbaša *et al.* 2011). Some authors have conducted kombucha fermentation with an amount of sucrose up to 10% (Blanc 1996).

Kombucha ferments under aerobic conditions at temperatures between 20°C and 30°C (Dufresne and Farnworth 2000; Malbaša *et al.* 2011; Jayabalan *et al.* 2014; Jayabalan, Malbaša, and Sathishkumar 2015, 2016). The vessels for kombucha fermentation should have a wide orifice to ensure active contact with oxygen in the air (Figure 8.1).

Substrate inoculation is possible using wet or dry kombucha pellicle and/or fermentative liquid from previous fermentations (Konovalov and Semenova 1955). Kombucha produces a cellulose pellicle as just mentioned, which is colored white to light brown (Figure 8.2).

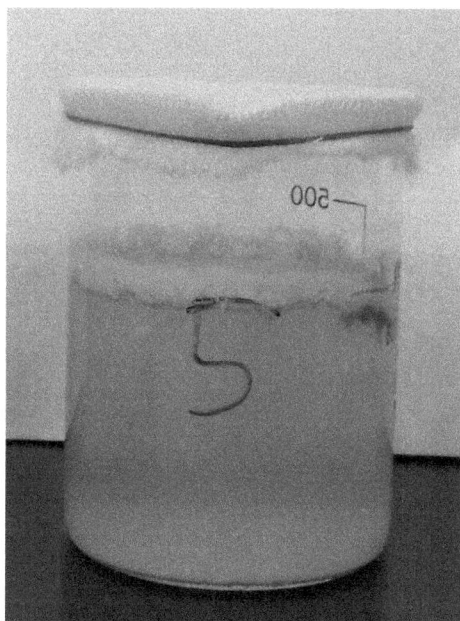

Figure 8.1   Kombucha Fermentation of Black Tea Sweetened with Sucrose in a Glass Beaker.

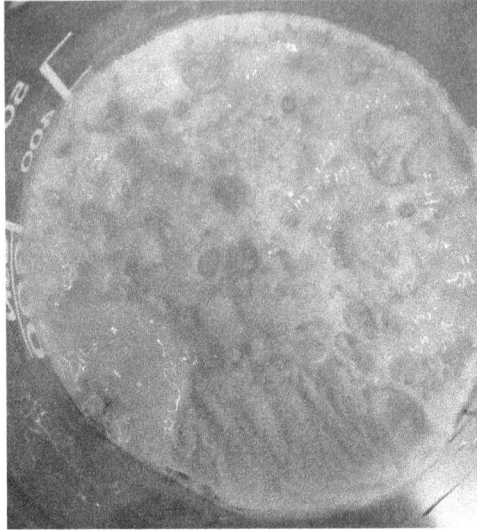

Figure 8.2   Cellulosic Pellicle Layer During Kombucha Fermentation of Black Tea Sweetened with Sucrose in a Glass Beaker.

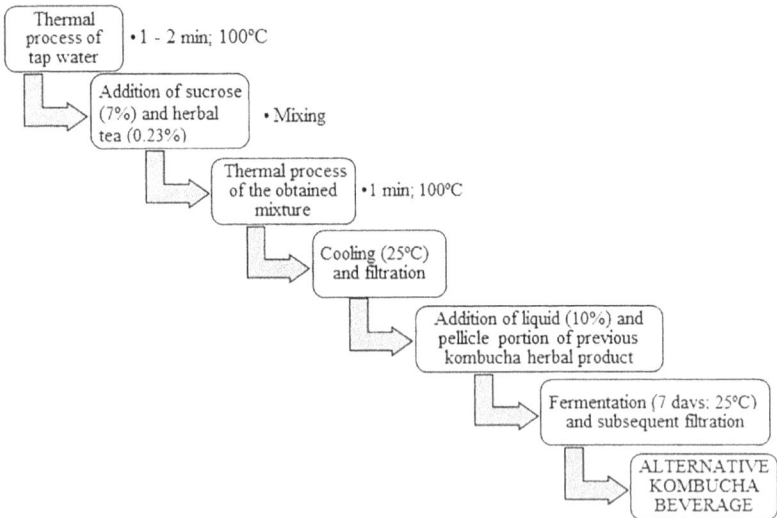

Figure 8.3   Example of the Basic Scheme of the Domestic or Laboratory Production of the Alternative Kombucha Beverage.

Source: Vitas (2013).

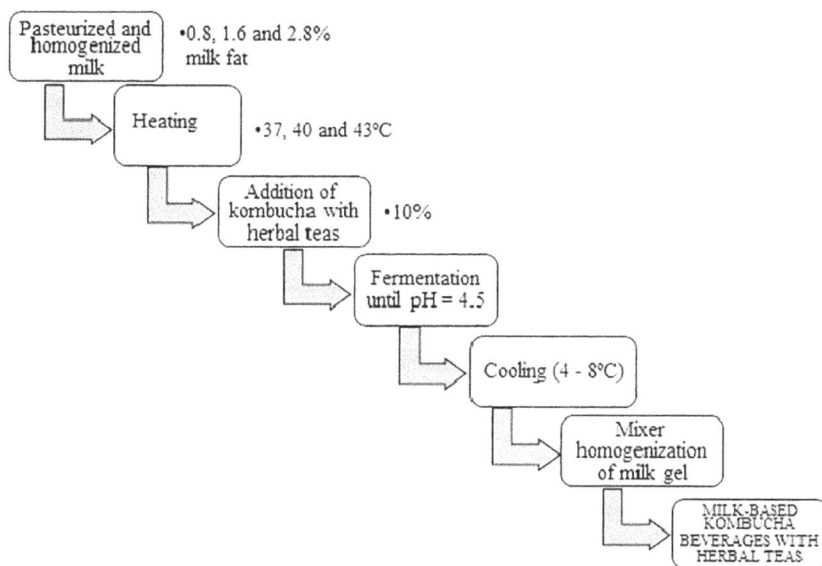

Figure 8.4 Example of the Basic Scheme for the Production of the Milk-Based Kombucha Beverages with Herbal Teas.

Source: Vitas (2013).

It floats on the surface of the fermentative liquid. The average duration of traditional kombucha fermentation is about seven days at room temperature. Kombucha products obtained by fermentation of any substrate other than black or green tea are known as alternative kombucha products. An example of the basic scheme of the domestic or laboratory production of the alternative kombucha beverage is presented in Figure 8.3.

Kombucha cultivation is possible on various alternative substrates and sources of carbon: beer, coffee, lactose, glucose, fructose, red wine, white wine, Jerusalem artichoke tubers extract, molasses, winter savory extract, peppermint extract, wild thyme extract, stinging nettle extract, quince extract, elderberry extract, winery effluent, milk, among others (Reiss 1994, 1987; Malbaša, Lončar, and Kolarov 2002; Vitas *et al.* 2013, 2020; Vukmanović, Vitas, and Malbaša 2020; Malbaša, Lončar, and Djurić 2008). Cultivation temperature on alternative substrates correlates with processing temperature for fermentation on the traditional substrate. The exception is kombucha fermentation on milk, which needs higher temperatures between 37°C and 43°C (Vitas *et al.* 2013). An example of the basic scheme for the production of milk-based kombucha beverages with herbal teas is presented in Figure 8.4.

8.3.1 Kombucha Microbiological Composition

Kombucha is the association of various osmophilic yeasts belonging to the genera *Schizosaccharomyces, Saccharomycodes, Saccharomyces, Zygosaccharomyces, Candida, Pichia, Kloeckera, Mycotorula, Mycoderma, Brettanomyces, Torulopsis,* and *Torulospora* (Mayser *et al.* 1995; Coton *et al.* 2017). Dominant yeast species in kombucha are given in Figure 8.5.

Identified species of the genus *Brettanomyces* include *Brettanomyces intermedius, B. bruxellensis,* and *B. claussenii.* The following species in the genus *Candida* were reported: *Candida famata, C. guilliermondii, C. obutsa, C. famata, C. stellate, C. guilliermondi, C. colleculosa, C. kefyr,* and *C. krusei.* Yeasts belonging to the *Saccharomyces* genus include *Saccharomyces cerevisiae* and *S. bisporus.* The reported species in the *Schizosacchromyces* genus include *Schizosaccharomyces pombe,* and among *Zygosaccharomyces, Zygosaccharomyces rouxii, Z. bailii,* and *Z. kombuchaensis* sp. n. were identified. Yeast species *Sacchromyccoides ludwigii* and *Schizosaccharomyces pombe* were also reported, apart from the previously mentioned species. The following yeast species were also reported: *Torula, Torulopsis, Torulaspora delbrueckii, Mycoderma, Pichia, Pichia membranefaciens, Kloeckera apiculata,* and *Kluyveromyces africanus.* No specific

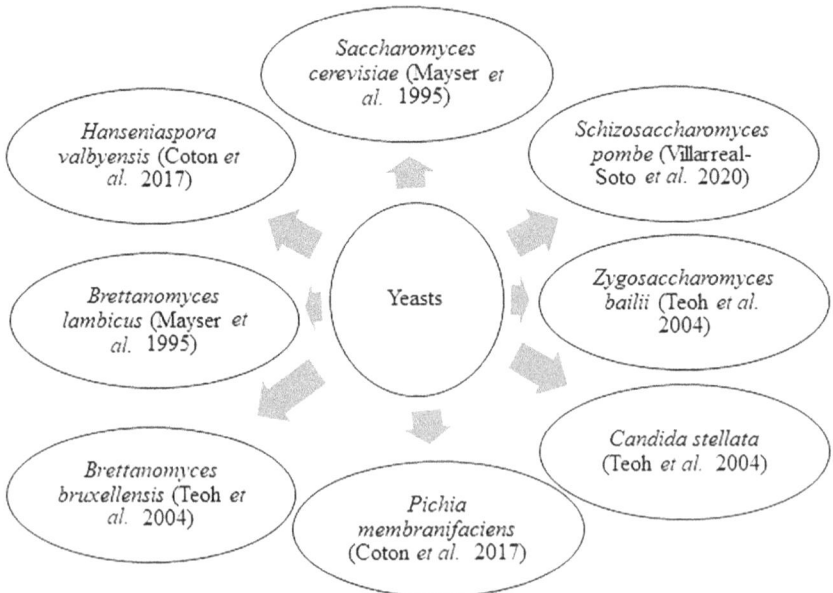

Figure 8.5   Dominant Yeast Species in Kombucha.

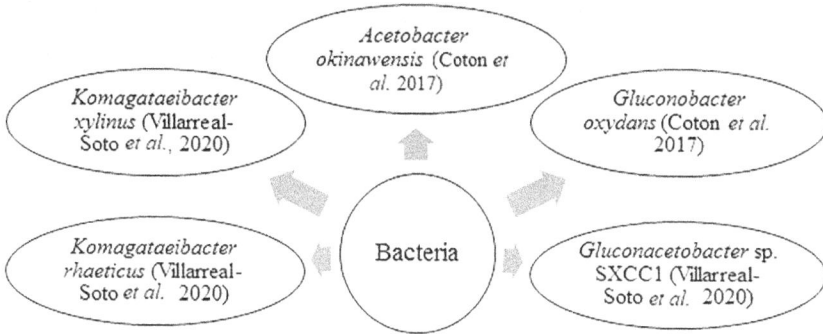

Figure 8.6   Dominant Bacteria Species in Kombucha.

yeast species was consistently found to be the main characteristic of the kombucha fermentation process. A range of acid-producing yeasts species, such
as *Brettanomyces* (*Dekkera*), are well adapted to acid environments, similar to
kombucha (Nguyen *et al.* 2010; Dufresne and Farnworth 2000; Teoh, Heard, and
Cox 2004; Jayabalan, Malbaša, and Sathishkumar 2015; Laureys, Britton, and
Clippeleer 2020).

Acetic acid bacteria (AAB) are the most characteristic microorganisms of
kombucha fermentation. They make up almost 80% of the microbes present in
the kombucha (Villarreal-Soto *et al.* 2020). The dominant bacteria species in
kombucha are presented in Figure 8.6.

There are 17 genera of AAB, but those found in kombucha belong to the
genera *Acetobacter*, *Gluconobacter*, *Gluconacetobacter*, and *Komagataeibacter*.
Among *Acetobacter*, *A. pasteurianus*, *A. aceti*, *A. intermedium* sp. nov., and *A.
nitrogenifigens* sp. nov. are found. Among *Gluconobacter* genus, *Gluconobacter
oxydans*, *Gluconoacetobacter* sp. A4 and *Gluconoacetobacter kombuchae* sp.
nov. are reported to be present. *Komagataeibacter xylinus* is thought to be
the most characteristic microorganism of kombucha fermentation and to be
responsible for producing the cellulose pellicle. This bacterium has been classified as *A. xylinus*, *A. xylinum*, *A. aceti* subsp. *xylinus*, *Gluconobacter xylinus*,
and *Gluconacetobacter xylinus*, but since 2012 it is classified in the separate
Komagataeibacter genus. *Komagateibacter* species can accumulate up to
10–20% acetic acid in the medium, while *Acetobacter* can accumulate only up
to 8% acetic acid. *Komagataeibacter rhaeticus* was found to make 16–49% of
all microbial population after 15 days of fermentation. In the pellicle, it was
followed by *Gluconobacter* sp. SXCCI, which makes up to 26% of the microbial
population. *Lactobacillus* species were found in kombucha recently, and their
presence is inconsistent. Usually, they are not present at all or present in low

amounts. Roussin (1996) tested around 900 kombucha samples and determined no presence of pathogenic *Candida albicans*, although some samples contained the molds *Aspergillus niger* and *Penicillium notatum* (Jayabalan, Malbaša, and Sathishkumar 2015; Laureys, Britton, and Clippeleer 2020; Yamada *et al.* 2012; Villarreal-Soto *et al.* 2020; Yamada, Yukphan, Vu, Muramatsu, Ochaikul, and Nakagawa 2012).

8.3.2 Kombucha Chemical Composition

The chemical composition of kombucha beverage is versatile and composed of sugars, acids, vitamins, enzymes, essential ions, alcohols, etc. (Malbaša 2004; Jayabalan *et al.* 2014).

In total, organic acids make 0.5–0.6% of the dry weight of the fresh tea. Acetic acid is the dominant acid in the kombucha beverage, with the concentration gradually increasing over time (Malbaša 2004).

Lactic acid is not always detected in kombucha beverage. Jayabalan, Marimuthu, and Swaminathan (2007) measured up to 0.44 g/L of lactic acid in black tea kombucha. The amount was the highest after three days of fermentation and after third day, it decreased. Vukmanović, Vitas, and Malbaša (2020) on the winery effluent–based kombucha measured around 0.1 g/L of lactic acid between days 0 and 4. It also decomposed over time.

Vitamin C is regularly detected in kombucha beverage, although the measured value varied greatly, from 4.4 mg/L to approximately 25 mg/L for black tea kombucha, with a continual increase over time. Milk-based kombucha with stinging nettle had around 50 mg/L, and kombucha made using winery effluent had maximal vitamin C content at the beginning of the fermentation (approximately 55 mg/L) with a decrease over time (Danielova 1957; Vitas *et al.* 2013; Vukmanović, Vitas, and Malbaša 2020; Malbaša *et al.* 2011).Bauer-Petrovska and Petrushevska-Tozi (2000) determined trace elements in kombucha. Manganese, iron, nickel, copper, and zinc increased in concentration after eight days of fermentation compared to tea decoct, while lead and chromium content decreased.

The ethanol content in kombucha was low, not surpassing 0.8% with a fermentation temperature of 20–30°C, while at 25°C, it was lower, up to 0.6% with wine effluent as the cultivation medium. Kombucha fermented on the traditional substrate had an ethanol content below 0.5% (Vitas *et al.* 2019; Blanc 1996; Malbaša *et al.* 2006; Reiss 1994). Vitas *et al.* (2021) demonstrated that it was possible to classify milk-based kombucha beverages with medicinal herbs taking into account their chemical composition, i.e., quality characteristics, using chemometric tools.

8.3.3 Kombucha Physiological Properties

Due to the high sugar content, the presence of a large number of organic acids and of water-soluble vitamins, kombucha beverage has clearly a high nutritional value. Various reports describe some of the healing properties of the beverage. Books and articles about kombucha mention diseases and physiological phenomena that can possibly be treated and prevented by kombucha beverage, such as headache, hemorrhoids, atherosclerosis by regeneration of cell walls, metabolic disorders, gout, arthritis, diabetes, intestinal laziness, insomnia, stress, obesity, hair loss, aging, hypertension, psoriasis, microbiological infections, cancer, weakened immune system, among others (Frank 1995; Sai Ram *et al.* 2000; Vijayaraghavan *et al.* 2000; Greenwalt, Steinkraus, and Ledford 2000; Sreeramulu, Zhu, and Knol 2000). The reported effects of kombucha from tea consumers' testimony and Russian researchers, besides the effects previously mentioned, are as follows (Dufresne and Farnworth 2000): detoxicating the blood, reducing cholesterol level, reducing inflammatory problems, alleviating rheumatism, promoting liver functions, regulating appetite, preventing and curing bladder infection and reducing kidney calcification, stimulating glandular systems, and having an antimicrobial effect against bacteria, viruses, and yeasts. Additionally, the reported effects include stimulating interferon production, relieving bronchitis and asthma, reducing menstrual disorders and menopausal hot flashes, reducing an alcoholic's craving for alcohol, and improving eyesight (Jayabalan *et al.* 2014). The authors agree that there is no solid scientific evidence for such claims. It is necessary to conduct comprehensive pharmacological tests, in which clinical studies would eventually furnish serious estimations on the healing properties of kombucha.

From a pharmacological viewpoint, the main kombucha metabolites are glucuronic acid, L-lactic acid, gluconic acid, and gluconates, as well as antioxidative vitamins. Glucuronic acid is involved in detoxication and prevention of oxidative stress, as well as prevention of skin wrinkles (Kaufman 1996). L-lactic acid is very important for cancer prevention (Kaufman 1996). Gluconates are very important for the absorption and transport of zinc, manganese (II), and chromium (III) ions in the body, which regulate the utilization of glucose, proteins, and lipids, as well as insulin secretion (Potter, Kies, and Rojhani 1990; Brun *et al.* 1995; Sun *et al.* 2000). Antioxidative vitamins, such as vitamin C, vitamin B2, and B6, are responsible for preventing oxidative stress and are quite certainly present in kombucha beverage (Jayabalan *et al.* 2014).

Kombucha shows antimicrobial activity toward various microorganisms, which has been known for a long time. Konovalov and Semenova (1955) determined that kombucha broth shows antimicrobial properties against *Staphylococcus aureus*. Danielova (1957) detected certain antimicrobial compounds she was unable to define. Hesseltine (1965) claimed that kombucha shows antibiotic properties against *Agrobacterium tumefaciens*. Antibiotic activity of kombucha beverage against *Helicobacter pylori*, *Agrobacterium tumefaciens*, *Escherichia coli*, and *Staphylococcus aureus* was determined. Certain growth inhibition against *Bacillus cereus* and *Salmonella enterica* serotype *typhimurium* was observed. Acetic acid was found to be the main antimicrobial compound, and no inhibition zone was observed in neutralized samples (Steinkraus *et al.* 1996; Greenwalt, Ledford, and Steinkraus 1998).

So far, the most comprehensive and complete research on the antimicrobial properties of kombucha beverage was carried out by Sreeramulu, Zhu, and Knol (2000). The activity of kombucha beverage against numerous pathogenic microorganisms, such as *Staphylococcus aureus*, *Shigella sonnei*, *Escherichia coli*, *Aeromonas hydrophilia*, *Yersinia enterocolitica*, *Pseudomonas aeruginosa*, *Enterobacter cloacae*, *Staphylococcus epidermis*, *Campylobacter jejuni*, *Salmonella enteritidis*, *Salmonella typhimurium*, *Bacillus cereus*, *Helicobacter pylori*, and *Listeria monocytogenes* was confirmed. Sreeramulu, Zhu, and Knol (2000) expanded the findings previously obtained by some authors (Steinkraus *et al.* 1996; Greenwalt, Ledford, and Steinkraus 1998), who concluded that only organic acids were responsible for the antimicrobial effects of kombucha beverage. After neutralizing and heating the kombucha beverage to 80°C, Sreeramulu, Zhu, and Knol (2000) proved that it exhibited activity against *Shigella sonnei*, *Escherichia coli*, *Campylobacter jejuni*, *Salmonella enteritidis*, and *Salmonella typhimurium*. This finding suggested that antimicrobial components, other than acetic acid or large protein molecules, are present in the kombucha beverage.

Dufresne and Farnworth (2000) have compared the biological activity of black tea and kombucha beverage in the human body. They reported that black tea lowers cholesterol, reduces atherosclerosis, lowers blood pressure, reduces inflammation, stimulates liver function, protects against diabetes, increases resistance to cancer, and has an antimicrobial effect against bacteria, viruses, and yeasts. Kombucha beverage showed all the effects of black tea, as well as the others previously mentioned. The reason for the different physiological effects of black tea and kombucha beverage was indisputably the different chemical composition. The dominant components of black tea are catechins, flavonoids, and methylxanthines, and they are most responsible for its physiological effects. Kombucha beverage contains all or only some of these compounds. In addition, it has other biologically very important compounds, such as glucuronic, gluconic, and L-lactic acid, water-soluble vitamins, and other compounds that do not have biological activity but that positively affect the action of biologically active components. This fact especially relates to many organic acids, which are very important for good sensory acceptability and for beverage preservation.

The presence of sucrose and free glucose and fructose, which are the dominant energy source of kombucha beverage, which are easily resorbed by the body and regulate appetite, and which reduce the feeling of weakness and similar ailments cannot be neglected.

8.4 VOLATILE COMPOUNDS AND FORMATION OF VOLATILE COMPOUNDS IN KOMBUCHA

There are different types of kombucha products, and fermentation is finished within various time periods. The volatile compounds' composition and content are the most important at consumption time. If fermentation progress needs to be evaluated, volatile compounds can be monitored at different time frames.

The AAB perform the periplasmatic oxidation of sugars, sugar alcohols, aldehydes, and alcohols in the presence of oxygen by dehydrogenases located on the outer surface of the cytoplasmic membrane. Ethanol is oxidized into acetaldehyde, which is further oxidized into acetic acid. *Acetobacter* and *Gluconacetobacter* species oxidize ethanol rather than glucose, while *Gluconobacter* species prefer the oxidation of glucose, glycerol, gluconic acid, and sorbitol over ethanol (Laureys, Britton, and Clippeleer 2020).

The predominant volatile metabolic product present in kombucha beverage is acetic acid. New research findings have given more detailed insight into the other volatile compounds of kombucha beverages.

8.4.1 Acetic Acid and Ethanol Formation in Kombucha Beverages

Acetic acid is produced and excreted by acetic acid bacteria, mostly genus *Acetobacter*, present in the kombucha culture (Figures 8.7 and 8.8).

Ethanol can be produced by fermentation of sugars by yeast, present in kombucha (Figure 8.7). Ethanol is not the characteristic kombucha metabolite because this fermentation process is aerobic, and AAB oxidize it further to acetic acid. Nevertheless, some quantities are always produced. In order to qualify kombucha beverages as nonalcoholic, ethanol levels must be under defined levels. The stipulated ethanol content in nonalcoholic beverages varies according to the regulations in countries around the world.

Carbon dioxide is produced by yeast degradation of sugars (Figure 8.7). To the best of our knowledge, the literature regarding carbon dioxide analysis in kombucha products is scarce. The CO_2 content in kombucha beverage was determined only by Sievers *et al.* (1995). The authors used the calculation method and established that CO_2 content was in the range from 4.3 (tenth day

```
        ┌─────────────────────┐
        │      Sucrose        │
        │  degradation  by    │
        │     invertase       │
        └─────────────────────┘
                  │
        ┌─────────────────────┐
        │   Glucose  and      │
        │     fructose        │
        └─────────────────────┘
                  │
          ┌───────┴────────┐
┌──────────────────┐  ┌──────────────────────┐
│                  │  │  Komagateibacter     │
│      Yeasts      │  │     xylinum          │
│                  │  │                      │
└──────────────────┘  └──────────────────────┘
          │                    │
┌──────────────────┐  ┌──────────────────────┐
│                  │  │                      │
│ Ethanol  and CO₂ │  │  Cellulosic  pellicle│
│                  │  │                      │
└──────────────────┘  └──────────────────────┘
          │
┌──────────────────┐
│  Acetobacter and │
│ Gluconacetobacter│
│ oxidation  of ethanol│
│  into  acetaldehyde │
└──────────────────┘
          │
┌──────────────────┐
│   Oxidation  of  │
│ acetaldehyde  into│
│    acetic acid   │
└──────────────────┘
```

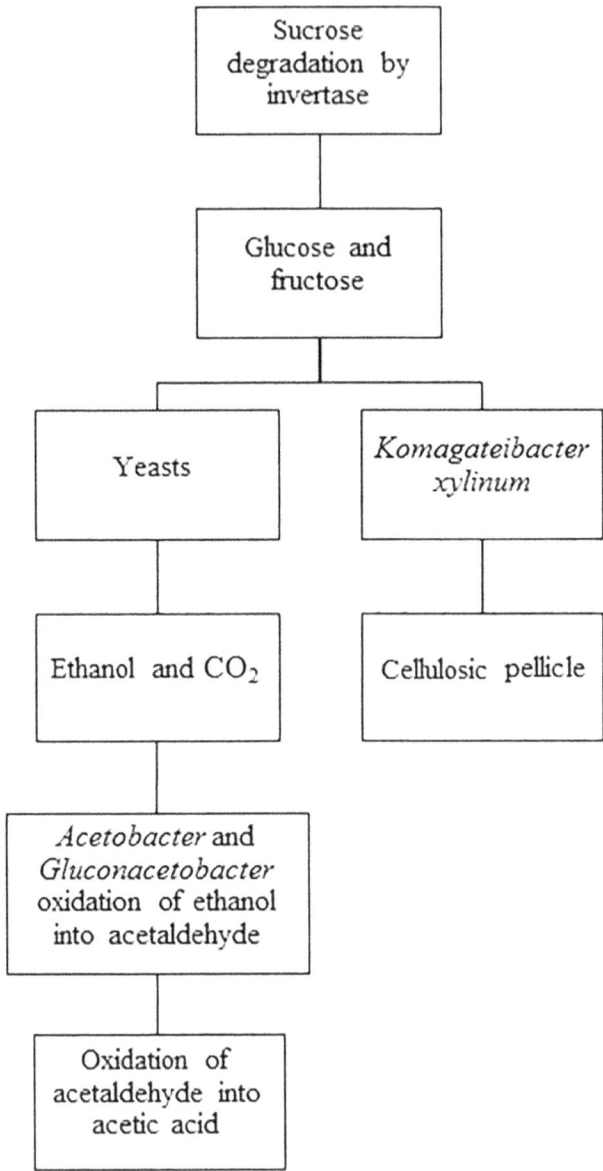

Figure 8.7   Updated Acetic Acid, Ethanol, and Carbon Dioxide Formation During Kombucha Fermentation.

Source: Based on the investigation by Sievers *et al.* (1995).

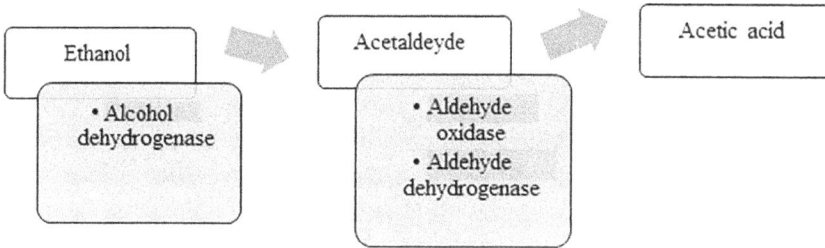

Figure 8.8   Aerobic Oxidation of Ethanol.

Source: Based on Karlson (1988).

of fermentation) to 20.2 g/L (40th day of fermentation). It was not detected at the beginning of the process.

Shahbazi *et al.* (2018) determined the acetic acid content in kombucha products with cardamom, Shirazi thyme, and cinnamon. Kombucha 4% v/v was added in the medium that contained 6.5% sucrose at 28°C for 16 days. Acetic acid was the dominant compound in all examined samples, and the content increased during the fermentation process. The highest content (2.8 g/L) was established for kombucha product with cinnamon (100%) at the end of fermentation (day 16 of the process).

Vitas *et al.* (2019) examined ethanol production in kombucha (10% liquid inoculum) products obtained by floated white wine must effluent fermentation. The initial substrate for kombucha fermentation had 5% of total sugars. The production process was conducted at 30°C, 25°C, and 20°C. Ethanol content was measured using the volumetric method characteristic for ethanol determination in refreshing nonalcoholic beverages. Analysis was performed in the initial substrate at the beginning of the fermentation and after three, six, and nine days of the process. The obtained results suggested that the product obtained after six days of fermentation at 30°C and samples produced after six and nine days at 20°C were not nonalcoholic beverages because they had ethanol content in an amount higher than 0.5% (v/v). The most suitable process temperature was 25°C because all of the obtained samples showed ethanol content lower than 0.5% (v/v).

Barbosa *et al.* (2020) determined the ethanol content in traditional black and green tea kombucha by the spectrophotometric method. Kombucha (10% liquid inoculum and 5% pellicle) was cultivated in a medium that contained 1.5% of tea and 8% sucrose. The fermentation process was carried out at 28°C for 15 days. At the end of the fermentation period (15 days), ethanol content was lower than 1% in all products. This study confirmed that traditional kombucha is a nonalcoholic beverage.

Talebi *et al.* (2017) analyzed the ethanol content of different commercial kombucha products. In this paper, the legal regulation by which kombucha products can be classified as nonalcoholic beverages stated that the highest amount was 0.5% (v/v) ethanol. Above this value, the examined product is an alcoholic beverage. Talebi *et al.* 2017 developed an HS-GC (head space–gas chromatography) method in order to determine the ethanol content. The investigation included a storage period at 4°C and 22°C. The chromatogram obtained of the analyzed sample showed the successful separation of ethanol from acetaldehyde, acetic acid, and unidentified volatile compounds. All analyzed kombucha samples were classified as alcoholic beverages. During the storage period, the ethanol content of analyzed samples increased. Similar trends were established for samples kept at 4°C, for closed and opened products. At 22°C, the ethanol content of closed kombucha beverages significantly increased after the seventh day of storage. A slight decrease was established after the 21st day, and the measured content remained approximately the same after 60 days. Talebi *et al.* (2017) suggested that some authors found that carbon dioxide production in closed kombucha containers can inhibit ethanol transformation to acetic acid. Refrigeration of products caused a reduction in ethanol production. Talebi *et al.* (2017) explained that the increase in ethanol production could result from ethyl esters hydrolysis. Also, it is possible that yeast activity related to the high content of sugar in kombucha can lead to an increase in ethanol content (Talebi *et al.* 2017). Rahmani *et al.* (2019) studied the ethanol and acetic acid content of African mustard kombucha beverage. The product was obtained by kombucha (2% liquid inoculum and around 7% pellicle) cultivation on a medium with 7% sucrose and around 1% of powdered African mustard leaves. The fermentation process lasted for 14 days at 25°C. HPLC (high-performance liquid chromatography) analysis was employed. The results of the measurements during the first ten days of the process showed an increase in content in both metabolites. Afterward, acetic acid content continued to increase, but the ethanol content decrease was significant and amounted to approximately 0.1% at the end of the process. These results were attributed to the fact that acetic acid bacteria metabolized ethanol into acetic acid, which is the main metabolic pathway of kombucha fermentation. The highest acetic acid content was measured at the ends of the process and was approximately 2.6%.

Zhang (2019) analyzed volatile acids, ethanol, and other volatile compounds in different kombucha products. These products were traditional (with black and green tea) and alternative (with white tea, chrysanthemum, honeysuckle, and mint). The fermentation medium contained 5% of sugar and 1% of the appropriate tea substrate. Kombucha was added in the amount of 10% liquid inoculum with a pellicle. The fermentation process was conducted at 25°C for 12 days. Acids and alcohols were determined using HPLC with UV (ultra-violet) and RI (refractive index) detection. In all examined beverages, acetic acid content did not differ significantly after six and 12 days of fermentation. Considering daily beverage consumption, the highest acetic acid content was established in honeysuckle kombucha (3.4 g/L after six days) and the lowest in mint kombucha (1.29 g/L after six days).

Tran *et al.* (2020) determined the acetic acid and ethanol content in traditional black tea kombucha (12% liquid inoculum) cultivated in a medium containing around 0.1% black tea and 6% sucrose. The products were obtained at 26°C. HPLC analysis was performed on samples after ten days in a closed vessel and after 14 days in an open vessel. The content of acetic acid, calculated as the difference between the end and the start of the fermentation process, was 1.6 and 2.1 g/L, respectively. In the same samples, using the enzymatic kit, the ethanol content difference was around 0.9 and 1.6 g/L, respectively.

Tan *et al.* (2020) produced kombucha (10% v/v) beverage using soursop (*Annona muricata* L.) as the substrate for the fermentation process with 5%, 10%, and 15% sugar. The fermentation process was performed at 28°C for 21 days. Among the volatile compounds, the content of acetic acid and ethanol was determined. The method of analysis was proton nuclear magnetic resonance metabolomics. Analysis was performed during the storage period of 21 days, and samples were analyzed on the 7th, 14th, and 21st days. Storage conditions were under light and in the dark, at room (25°C) and chilled temperature (4°C). The influence of dark and light storage conditions and the temperature had a minor effect on the acetic acid content. The exception was the sample on the 21st day of storage in low temperature, which contained the lowest acetic acid content (0.07 mmol/L). This value was significantly different from all the others and can be explained by the reduced acetic acid bacteria activity during prolonged storage at low temperatures. The highest value (approximately 0.64 mmol/L) was measured after 21 days of storage at 25°C. This result can be explained by the increased metabolic activity of acetic acid bacteria during the prolonged storage period at room temperature. The ethanol content was the highest at the start of the storage period and amounted to 3.28 mmol/L (Tan, Muhialdin, and Meor Hussin 2020). During the storage, the ethanol content significantly decreased. On day 7, light and temperature had no impact on the content. On the 14th day of storage, light and temperature showed some impact on the ethanol content. The impact of light was more pronounced for samples stored at 25°C. On day 21, the temperature of 4°C was more suitable for ethanol production. The lowest ethanol content (0.06 mmol/L) was measured on the 21st day of storage for the sample at room temperature in the dark (Tan, Muhialdin, and Meor Hussin 2020).

8.4.2 Other Volatiles Formation in Kombucha Beverages

It is important to emphasize that volatile compounds content is related to the kombucha fermentation process. By influencing the fermentation parameters, it is possible to determine the formation of the volatile compounds (Zhao *et al.* 2018).

Zhao *et al.* (2018) analyzed the volatile flavor compounds composition of kombucha products during a 14-day fermentation period at 28°C. Kombucha (2% liquid inoculum) was cultivated on the medium that contained 10% sucrose and 0.8% of raw Pu-erh tea. Volatile flavor compounds were investigated in Pu-erh tea kombucha before and during the fermentation, after 0, three, five, seven, ten, and 14 days of fermentation. The applied analysis method was GC-MS (gas chromatography–mass spectrometry) of HS-SPME (head space–solid-phase microextraction) extracts. Analyzed compounds were categorized as benzenoids, esters, alcohols, acids, ketones, aldehydes, among others. The obtained results indicated that the kombucha fermentation process influenced the production of volatile flavor compounds by approximately 2.7 times. Nearly 60% of all determined compounds were acids.

Fermentation time influenced volatiles composition, and the number of determined compounds increased during the process, reached the maximum on the seventh day, and slightly decreased on the tenth and 14th days. The content was expressed as relative in peak area units. Qualitative and quantitative composition depended on the fermentation process dynamics (Zhao *et al.* 2018).

At the start of fermentation, the majority of components came from the initial medium used for the fermentation, i.e., the Pu-erh tea. The abundance percentage of the dominant volatile compounds during fermentation is presented in Figure 8.9.

The highest content at the beginning of fermentation was measured for benzeneacetaldehyde and linalool. After three days, the highest values were determined for heptaldehyde and benzeneacetaldehyde. Acetic acid was the predominant volatile compound until the end of the process.

Zhao *et al.* (2018) classified the kombucha flavor compounds based on their origin. These compounds are as follows: nitrogen and carbon atom sources (mostly tea substrate used for fermentation, with added sugar) and numerous products of kombucha metabolism. The volatile compounds that originated from the Pu-erh tea used as a substrate for kombucha fermentation contributed the most to the flavor of the obtained beverage in the first part of the process (up to three days). The fruity aroma of the kombucha beverage resulted from benzeneacetaldehyde, 2,4-dimethyl benzaldehyde, heptaldehyde, and linalool presence. During the fermentation, these compounds were decomposed as the microbial population changed. Acetic acid was the predominant

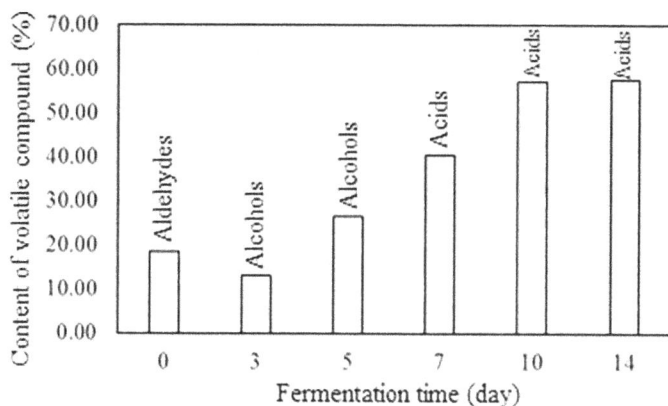

Figure 8.9 Percentage Contribution of the Dominant Volatile Compounds During Fermentation Time.

Source: Based on results of Zhao *et al.* (2018).

volatile compound that showed an increase in content during the fermentation process. Results suggested that during the fermentation, many biochemical reactions occur that cause the production of aldehydes and alcohols and their transformation to acids and esters.

Zhao *et al.* (2018) concluded that volatile flavor components changed dynamically during the kombucha fermentation. The dominant compounds belonged to acids, aldehydes, alcohols, and esters, and these were produced mostly after five days of fermentation.

For identifying volatile compounds in water and organic solvent extracts of African mustard kombucha and unfermented African mustard, GC-MS analysis, with prior derivatization of samples, was used (Rahmani *et al.* 2019). Acids and alcohol (glycerol) were identified in examined samples. 2-Hydroxy-3-methylbutyric and butanedioic acid were not detected in unfermented fractions of samples. The results indicated that kombucha fermentation led to the production of volatile compounds, such as acids and alcohol (Rahmani *et al.* 2019).

By Zhang (2019) other volatile compounds were extracted with HS-SPME, and GC-MS detection was employed for the analysis. Zhang (2019) determined that isovaleric acid was the dominant metabolite in most of the kombuchas with tea. The highest content (4.01 g/L) was determined in black tea kombucha with SCOBY2 on the 12th day of fermentation. Its content was higher in comparison to the kombuchas with herbs. Zhang (2019) concluded that the acidity of the kombucha products with tea created a taste that was less pleasant in comparison to the taste of the kombucha beverages with herbs. This was attributed to the isovaleric acid.

Zhang (2019) analyzed volatile compounds of traditional and herb-based kombucha and established the profile of alcohols, esters, fatty acids, ketones, aldehydes, and other components. Analysis was performed at the start of the fermentation, after six days, and after 12 days. Acids were determined in all analyzed samples.

Among volatile compounds, aroma carriers were determined. These components include phenyl ethyl alcohol (lilac aroma), *p*-ethylguaiacol (spice aroma), and ethyl octanoate (apricot aroma). Zhang (2019) established that a difference in the origin of kombucha starters did not influence the aroma profile of black tea kombucha. Significant changes in aroma profile were determined for different types of tea: black, white, and green. Kombucha with white tea contained *p*-methylanisate and neodihydrocarveol, which were not determined in kombucha with the black and green teas.

Zhang (2019) determined that kombuchas with herbs contained 17 components more in comparison to kombuchas with tea. Eugenol was determined only in kombucha with mint and chrysanthemum. Pinocarveol, *trans*-carveol, *cis*-carveol, and *p*-cymene were determined in kombucha with mint and honeysuckle. Jasmone, piperitenone, dihydro-carvone, dihydro-carveol, (-)-*trans*-dihydrocarvyl acetate, alpha-cadinol, and 3-octanol were determined only in kombucha with mint. These results suggested that the kombucha fermentation process did not influence the usual aroma composition of mint.

The main conclusion of Zhang (2019) was that aroma compounds characteristic of the substrates used for kombucha fermentation were retained after the process, and characteristic kombucha aromas were also produced. These include esters, alcohols, and acids.

Principal component analysis (PCA), as a statistical tool, suggested that the aroma profile of kombucha with various types of tea and chrysanthemum did not differ significantly. In contrast, honeysuckle and mint kombucha had per se characteristic aroma compounds.

Malbaša, Lončar, and Kolarov (2002) examined the volatile acids in dietetic kombucha beverage obtained by kombucha fermentation of Jerusalem artichoke tuber. Kombucha was cultivated on a medium that contained 0.15% of Indian black tea. The fermentation process lasted for 21 days. Volatile acids content, expressed as g/L, was higher in Jerusalem artichoke tuber kombucha beverage when compared to black tea kombucha. These volatiles were determined by the volumetric method.

Tu *et al.* (2019) determined that kombucha culture (10% v/v) can be successfully applied in the transformation of soy whey into a novel type of functional beverage. The fermentation process was performed at 28°C for eight days. As the result of this fermentation process, new volatile compounds (higher aldehydes and esters) responsible for fruity flavor were produced. Volatile compounds were analyzed using HS-SPME GC-MS and flame ionization detection (FID). Acetic acid was determined by GC, but also by using an HPLC method. HPLC analysis of acetic acid indicated that this compound is the main organic acid during fermentation. The content showed a linear increase during the fermentation time, and the highest value was established on the sixth day (5.79 g/L) (Tu *et al.* 2019). As the result of the GC method, the following three groups of volatile compounds were identified in kombucha fermented soy whey products: acids, aldehydes, and esters. Also, one alcohol was identified as well, 2-ethyl-1-dodecanol, but in a very small amount. Analysis was performed on the sample from day 6 because this product obtained the highest sensory score. The content of volatiles present in the soy whey was reduced to very low or undetectable levels, which caused the disappearance of the unpleasant beany flavor of soy. On the other hand, new aroma compounds (mainly aldehydes and esters) were produced and influenced the new floral and fruity flavor of the products (Tu *et al.* 2019). Nonanal and undecanal caused the pleasant orange and rose smell of the new beverage. Undecanal was produced in a significantly higher amount compared to nonanal, and production of these two aldehydes was stimulated by kombucha activity. Aldehyde *cis,cis*-7,10-hexadecadienal was also determined in the fermented product. Two acids (acetic and hexanoic) were measured. The content of acetic acid was around six times higher in comparison to the content of hexanoic acid. Results obtained for alcohols indicated that these compounds were transformed into acids and other substances. It was concluded that hexanoic acid was produced from 1-hexanol (responsible for the unpleasant flavor of the soy whey) (Tu *et al.* 2019). Among the esters, ethyl esters were predominant, and the presence of ethanol explained this. The determined esters were the following: hexadecanoic acid, ethyl ester, dodecanoic acid, ethyl ester, tetradecanoic acid, ethyl ester, phthalic acid, hex-3-yl isobutyl ester, and m-toluic acid, 2-ethylhexyl ester (Tu *et al.* 2019).

8.5 CONCLUSION

Kombucha beverage has been known for a very long time, but its popularity has significantly increased in the recent years. Besides the traditional form, there are a number of alternative kombucha beverages as well. They can be obtained, for example, by kombucha fermentation of winter savory, wild thyme, peppermint, stinging nettle, soy whey, milk, Jerusalem artichoke tuber, African mustard, chrysanthemum, honeysuckle, wine effluent, soursop, cardamom, cinnamon, among other ingredients. Volatile compounds are included in the chemical composition of kombucha beverages. They can originate from the substrate that was used for fermentation, or they can be produced during the process. The main compound is acetic acid. Besides other acids, esters, aldehydes, and ketones can be produced as well. In general, ethanol and carbon dioxide are regarded as minor volatile kombucha metabolites.

REFERENCES

Barbosa, C. D., M. R. Baqueta, W. C. Rodrigues Santos *et al.* 2020. Data fusion of UPLC data, NIR spectra and physicochemical parameters with chemometrics as an alternative to evaluating kombucha fermentation. *LWT* 133:109875.

Bauer-Petrovska, B., and L. Petrushevska-Tozi. 2000. Mineral and water soluble vitamin content in the kombucha drink. *International Journal of Food Science and Technology* 35:201–205.

Blanc, P. 1996. Characterization of the tea fungus metabolites. *Biotechnology Letters* 18:139–142.

Brun, J. F., R. Guintrand-Hugret, C. Fons *et al.* 1995. Effects of oral zinc gluconate on glucose effectiveness and insulin sensitivity in humans. *Biological Trace Element Research* 47:385–391.

Coton, M., A. Pawtowski, B. Taminiau *et al.* 2017. Unraveling microbial ecology of industrial-scale kombucha fermentations by metabarcoding and culture-based methods. *FEMS Microbiology Ecology* 93:1–16.

Danielova, L. T. 1957. K Himicheskomu sostavu i fiziko-khemicheskim svoistvam kulturalnoi zhidkosti chainoga gryba. *Trudy Erevanskoga Zooveterinarnogo Instituta* 22:111–121.

Dufresne, C., and E. Farnworth. 2000. Tea, kombucha, and health: A review. *Food Research International* 33:409–421.

Filho, L. X., M. Q. Paulo, E. C. Pereira, and C. Vicente. 1985. Phenolics from tea fungus analyzed by high performance liquid chromatography. Фyton 45:187–191.

Frank, G. W. 1995. *Das Teepliz-Getränk*. Steyr, Austria: Ennsthaler Verlag.

Greenwalt, C. J., R. A. Ledford, and K. H. Steinkraus. 1998. Determination and characterization of the antimicrobial activity of the fermented tea kombucha. *LWT—Food Science and Technology* 31:291–296.

Greenwalt, C. J., K. H. Steinkraus, and R. A. Ledford. 2000. Kombucha, the fermented tea: Microbiology, composition, and claimed health effects. *Journal of Food Protection* 63:976–981.

Hesseltine, C. W. 1965. A millennium of fungi, food, and fermentation. *Mycologia* 57:149–197.

Jayabalan, R., R. V. Malbaša, E. S. Lončar, J. S. Vitas, and M. Sathishkumar. 2014. A review on kombucha tea—microbiology, composition, fermentation, beneficial effects, toxicity, and tea fungus. *Comprehensive Reviews in Food Science and Food Safety* 13:538–550.

Jayabalan, R., R. V. Malbaša, and M. Sathishkumar. 2015. Kombucha tea: Metabolites. In *Fungal Metabolites: Reference Series in Phytochemistry*, edited by J. M. Mérillon and K. G. Ramawat, 965–978. Cham: Springer.

Jayabalan, R., R. V. Malbaša, and M. Sathishkumar. 2016. Kombucha. *Reference Module in Food Science*: 1–8.

Jayabalan, R., S. Marimuthu, and K. Swaminathan. 2007. Changes in content of organic acids and tea polyphenols during kombucha tea fermentation. *Food Chemistry* 102:392–398.

Karlson, P. 1988. *Kurzer Lehrbuch Der Biochemie Für Mediziner Und Naturwissenchaftler*, 13. Auflag. Stuttgart, Germany: Georg Thieme Verlag.

Kaufman, K. 1996. *Kombucha Rediscovered!* Burnaby, BC: Books Alive.

Konovalov, I. N., and M. N. Semenova. 1955. K Fiziologii "Chainoga Griba". *Botanicheskii Zhurnal (Moscow & Leningrad)* 40:567–570.

Laureys, D., S. J. Britton, and J. De Clippeleer. 2020. Kombucha tea fermentation : A review. *Journal of the American Society of Brewing Chemists* 78:165–174.

Lindau, G. 1913. Über medusomyces gisevil, eine neue gattung und art der hefepilze. *Berichte Der Deutschen Botanischen Gesellschaft* 31:243–248.

Maarse, H. 1991. *Volatile Compounds in Foods and Beverages*. Edited by Henk Maarse. New York: Marcel Dekker, Inc.

Malbaša, R. V. 2004. *Istraživanje Antioksidativne Aktivnosti Napitka Od Čajne Gljive*. Novi Sad: University of Novi Sad, Faculty of Technology Novi Sad.

Malbaša, R. V., E. Lončar, and M. Djurić. 2008. Comparison of the products of kombucha fermentation on sucrose and molasses. *Food Chemistry* 106:1039–1045.

Malbaša, R. V., E. Lončar, M. Djurić, M. Klašnja, L. J. Kolarov, and S. Markov. 2006. Scale-up of black tea batch fermentation by kombucha. *Food and Bioproducts Processing* 84:193–199.

Malbaša, R. V., E. S. Lončar, and LJ. A. Kolarov. 2002. L-Lactic, L-Ascorbic, total and volatile acids contents in dietetic kombucha beverage. *Roumanian Biotechnological Letters* 7:891–895.

Malbaša, R. V., E. S. Lončar, J. S. Vitas, and J. M. Čanadanović-Brunet. 2011. Influence of starter cultures on the antioxidant activity of kombucha beverage. *Food Chemistry* 127:1727–1731.

Mayser, P., S. Fromme, G. Leitzmann, and K. Gründer. 1995. the yeast spectrum of the "tea fungus kombucha": Das hefespektrum des "teepilzes kombucha". *Mycoses* 38:289–295.

Nguyen, V. T., B. Flanagan, D. Mikkelsen *et al.* 2010. Spontaneous mutation results in lower cellulose production by a gluconacetobacter xylinus strain from kombucha. *Carbohydrate Polymers* 80:337–343.

Potter, S. M., C. V. Kies, and A. Rojhani. 1990. Protein and fat utilization by humans as affected by calcium phosphate, calcium carbonate, and manganese gluconate supplements. *Nutrition* 6:309–312.

Rahmani, R., S. Beaufort, S. A. Villarreal-Soto, P. Taillandier, J. Bouajila, and M. Debouba. 2019. Kombucha fermentation of African mustard (Brassica Tournefortii) leaves: Chemical composition and bioactivity. *Food Bioscience* 30:100414.

Reiss, J. 1987. Der Teepiltz Und Sein Stoffwechselprodukte. *Deutche Lebensmittel-Rundschau* 83:286–290.

Reiss, J. 1994. Influence of different sugars on the metabolism of the tea fungus. *Zeitschrift Für Lebensmittel-Untersuchung Und -Forschung* 198:258–261.

Roussin, M. R. 1996. *Analyses of kombucha ferments*. Salt Lake City, UT: Information Resources, LC.

Sai Ram, M., B. Anju, T. Pauline *et al.* 2000. Effect of kombucha tea on chromate(vi)-induced oxidative stress in albino rats. *Journal of Ethnopharmacology* 71:235–240.

Shahbazi, H., H. H. Gahruie, M. T. Golmakani, M. H. Eskandari, and M. Movahedi. 2018. Effect of medicinal plant type and concentration on physicochemical, antioxidant, antimicrobial, and sensorial properties of kombucha. *Food Science and Nutrition* 6:2568–2577.

Sievers, M., C. Lanini, A. Weber, U. Schuler-Schmid, and M. Teuber. 1995. Microbiology and fermentation balance in a kombucha beverage obtained from a tea fungus fermentation. *Systematic and Applied Microbiology* 18:590–594.

Sreeramulu, G., Y. Zhu, and W. Knol. 2000. Kombucha fermentation and its antimicrobial activity. *Journal of Agricultural and Food Chemistry* 48:2589–2594.

Steinkraus, K. H., K. B. Shapiro, J. H. Hotchkiss, and R. P. Mortlock. 1996. Investigations into the antibiotic activity of tea fungus/kombucha beverage. *Acta Biotechnologica* 16:199–205.

Sun, C., W. Zhang, S. Wang, and Y. Zhang. 2000. Effect of chromium gluconate on body weight, serum leptin and insulin in rats. *Wei Sheng Yan Jiu* 29:370–371.

Talebi, M., L. A. Frink, R. A. Patil, and D. W. Armstrong. 2017. Examination of the varied and changing ethanol content of commercial kombucha products. *Food Analytical Methods* 10:4062–4067.

Tan, W. C., B. J. Muhialdin, and A. S. Meor Hussin. 2020. Influence of storage conditions on the quality, metabolites, and biological activity of soursop (Annona Muricata. L.) Kombucha. *Frontiers in Microbiology* 11:1–10.

Teoh, A. L., G. Heard, and J. Cox. 2004. Yeast ecology of kombucha fermentation. *International Journal of Food Microbiology* 95:119–126.

Tran, T., C. Grandvalet, F. Verdier, A. Martin, H. Alexandre, and R. Tourdot-Maréchal. 2020. Microbial dynamics between yeasts and acetic acid bacteria in kombucha: Impacts on the chemical composition of the beverage. *Foods* 9:963.

Tu, C., S. Tang, F. Azi, W. Hu, and M. Dong. 2019. Use of kombucha consortium to transform soy whey into a novel functional beverage. *Journal of Functional Foods* 52:81–89.

Vijayaraghavan, R., M. Singh, P. V. Rao *et al.* 2000. Subacute (90 days) oral toxicity studies of kombucha tea. *Biomedical and Environmental Sciences* 13:293–299.

Villarreal-Soto, S. A., J. Bouajila, M. Pace *et al.* 2020. Metabolome-microbiome signatures in the fermented beverage, kombucha. *International Journal of Food Microbiology* 333:108778.

Vitas, J. S. 2013. *Antioksidativna Aktivnost Fermentisanih Mlečnih Proizoda Dobijenih Pomoću Kombuhe.* Novi Sad: University of Novi Sad, Faculty of Tehnology Novi Sad.

Vitas, J. S., M. Karadžić Banjac, S. Kovačević *et al.* 2021. Chemometric approach to quality characterization of milk-based kombucha beverages. *Mljekarstvo* 71:83–94.

Vitas, J. S., R. V. Malbaša, J. A. Grahovac, and E. S. Lončar. 2013. The antioxidant activity of kombucha fermented milk product with stinging nettle and Winter savory. *Chemical Industry and Chemical Engineering Quarterly* 19:129–139.

Vitas, J. S., S. Vukmanović, J. Čakarević, L. Popović, and R. Malbaša. 2020. Kombucha fermentation of six medicinal herbs: Chemical profile and biological activity. *Chemical Industry and Chemical Engineering Quarterly* 26:157–170.

Vitas, J. S., S. Z. Vukmanović, R. V. Malbaša, and A. N. Tepić Horecki. 2019. Influence of process temperature on ethanol content in kombucha products obtained by fermentation of flotated must effluent. *Acta Periodica Technologica* 50:311–315.

Vukmanović, S., J. Vitas, and R. Malbaša. 2020. Valorization of winery effluent using kombucha culture. *Journal of Food Processing and Preservation* 44:e14627.

Yamada, Y., P. Yukphan, H. T. L. Vu, Y. Muramatsu, D. Ochaikul, and Y. Nakagawa. 2012. Subdivision of the genus gluconacetobacter Yamada, Hoshino and Ishikawa 1998: The proposal of komagatabacter gen. nov., for strains accommodated to the gluconacetobacter xylinus group in the α-proteobacteria. *Annals of Microbiology* 62:849–859.

Yamada, Y., P. Yukphan, H. T. L. Vu *et al.* 2012. Description of komagataeibacter gen. nov., with proposals of new combinations (acetobacteraceae). *Journal of General and Applied Microbiology* 58:397–404.

Zhang, J. 2019. *A Comprehensive Study on Kombucha and Its Analogues*. Lavras: The Federal University of Lavras.

Zhao, Z. J., Y. C. Sui, H. W. Wu, C. B. Zhou, X. C. Hu, and J. Zhang. 2018. Flavour chemical dynamics during fermentation of kombucha tea. *Emirates Journal of Food and Agriculture* 30:732–741.

Chapter 9

Volatile Compounds Formation in Dark Tea

Zisheng Han and Liang Zhang

CONTENTS

9.1 INTRODUCTION

Dark tea is a specific type of tea of southwestern China that is processed by fermenting with environmental or artificial microorganisms. Different from nonfermented or fully fermented teas, dark tea has its characteristic chemicals and flavor. During the processing of dark tea, the fresh tea leaves and stems are also first deactivated by high temperature and then piled in controlled conditions under high temperature and relative humidity for a few months or even a few years. Usually, the raw materials of most of dark tea are mature leaves and stems, so that the contents of astringent and bitter compounds are much higher than those of tea shoots or young leaves. After long-term postfermentation by microorganisms, the flavor of dark tea becomes less astringent and bitter.

Of the typical dark teas in China, some are geographically associated: for example, ripened pu-erh tea in Yunnan, fuzhuan tea in Hunan, Tibetan tea in Sichuan, and liubao tea in Guangxi of China. The degree of postfermentation determines the level of retained polyphenols, especially galloylated and nongalloylated catechins in processed teas. When comparing the catechins

DOI: 10.1201/9781003129462-11

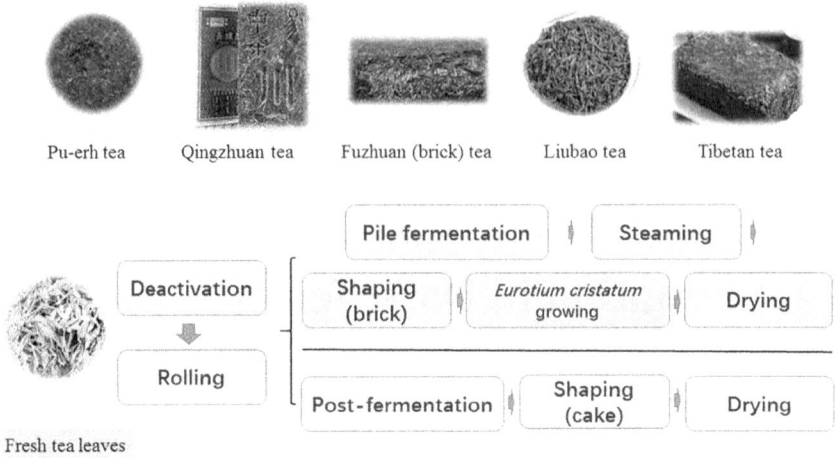

Figure 9.1   Typical Types and Major Processing Procedures of Dark Tea.

content of crude (green) pu-erh tea to that of postfermented, ripened pu-erh tea, it can be concluded that fermentation causes a decrease in the concentration of such compounds. Nevertheless, some novel catechins derivatives are formed during this stage. Except for ripened pu-erh tea, the degree of fermentation of other dark teas is relative mild so that the levels of the main polyphenols are still detected at 1–10 mg/g, such as in fuzhuan tea and liubao tea. Figure 9.1 shows the major types and production procedures of dark tea.

9.2 AROMA OF DARK TEA

The aroma of dark tea is very different from that of green, black, and oolong teas in which the floral, sweet, and faint aromas are easily perceived. Basically, the linalool, geraniol, linalool oxide, α-terpenol, α-ionone, and β-ionone are widely detected and recognized as critical aroma compounds of green teas. In the fresh tea leaves and processed teas, the biosynthesis of aroma compounds is derived from glycosidic aroma precursors, carotenoids, and long-chain fatty acids. The Maillard reaction of amino acids or unique and abundant L-theanine in *Camellia sinensis* and *Camellia assamica* is also involved in the formation of N-containing heterocyclic aromatic substances.

The aroma description of dark teas has a universal flavor perception that is aged flavor. In pu-erh tea, the main contributors of aged flavor contain 1,2,3-trimethoxy-5-methylbenzene, 4-ethyl-1,2-dimethoxybenzene,

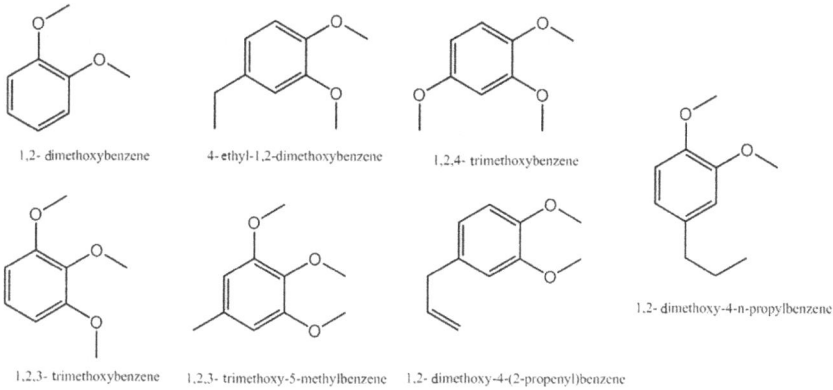

1,2- dimethoxybenzene   4- ethyl-1,2-dimethoxybenzene   1,2,4- trimethoxybenzene

1,2- dimethoxy-4-n-propylbenzene

1,2,3- trimethoxybenzene   1,2,3- trimethoxy-5-methylbenzene   1,2- dimethoxy-4-(2-propenyl)benzene

Figure 9.2   Chemical Structures of Some Methoxybenzenes.

1,2-dimethoxybenzene, 1,2,3-trimethoxybenzene, 1,2,4-trimethoxybenzene. It was supposed that these methoxybenzenes are degradation products of gal-loyl moiety of catechins or hydrolysable tannins, the contents of which are significantly decreased after postfermentation. The chemical structures of some methoxybenzenes are shown in Figure 9.2. Woody aroma is also an important characteristic of dark teas. Dihydroactinidiolide, α-guaiene, α-cedarene, α-cedarol, α-ionone, and β-ionone are recognized as critical volatile compounds for woody aroma. *Eurotium cristatum* is the dominant flora of fuzhuan tea processing. During postfermentation, which is the growing of *Eurotium cristatum*, the content levels of linalool oxide I, linalool oxide II, linalool, 1-octene-3-ol are significantly changed.

9.3 VOLATILE COMPOUNDS OF DARK TEA

Shi *et al.* (2019) studied the volatile compositions of fuzhuan (fu-brick) and pu-erh tea. Simultaneous distillation extraction was used to extract the volatile compounds, which were subsequently analyzed by two-dimensional gas chromatography (2D-GC) mass spectrometry (TOF-MS). Obviously, the 2D-GC had better separation efficiency and expanded the peak capacity. In total, 373 and 408 compounds, respectively, were detected and identified in fu-brick tea and pu-erh tea, and the ketenes, ketones, aldehydes presented at high concentrations. In the subclass of ketenes β and α-ionone are key aromatics for floral aroma in pu-erh tea and fuzhuan tea. We previously reported the increasing β-ionone during the beginning stage of tea leaves fermentation with *Aspergillus niger* and *Eurotium cristatum* (Cao *et al.* 2018). Furthermore, 1-octen-3-one was supposed to be produced during postfermentation by microorganisms. Du *et al.* (2019) also did the volatiles analysis and sensory evaluation for instant pu-erh tea by 2D-GC-MS, in which a total of 21 methoxy-phenolic compounds were identified. In this study, volatile compounds were extracted by SPME, and the results showed that hydrocarbons, ketones, and alcohols were the main volatile compounds. Another study also compared the aroma compounds extracted by solid-phase microextraction and simultaneous distillation-extraction (SDE). It suggested that SDE may collect high-molecular-weight alcohols, acids, and esters, while SPME can capture much more methoxy-phenolic compounds, which have been associated with the stale flavor of many dark teas. In both extraction methods, methoxy-phenolic compounds, aldehydes, and alcohols accounted for the main volatile compositions in decreasing order (%) (Du *et al.* 2014). A similar methodology research was also conducted on fuzhuan tea. The SDE extracted many more (79 kinds) of hydrocarbons, heterocyclic oxygens, and phenols (mainly methyl-phenol) compared with 50 kinds of compounds extracted by SPME (Li *et al.* 2018).

Because of the variation of dark tea samples, the process technology used, and the microorganism involved, the volatile composition may show differences. Lv *et al.* (2014) analyzed the volatile compositions of Dayi pu-erh tea, product from the largest pu-erh tea company in China by HS-SPME/GC-MS. Methoxyphenolic compounds (36.08%), hydrocarbons (14.54%), and alcohol compounds (13.58%) were the major aromatic components in the seven Dayi pu-erh tea samples, in which 1,2,3-trimethoxybenzene was identified as the most abundant compound for all Dayi pu-erh tea samples. They compared differences in volatiles between Dayi pu-erh tea and Ya'an dark tea, the famous dark tea in Sichuan province of China. The results showed that ketones (34.00%) and hydrocarbon compounds (17.38%) were major aromatic components in Ya'an dark teas. Furthermore, linalool oxides, α-terpineol, α-cadinol, isophytol, and phytol with typically floral and sweet scents were also detected in Dayi pu-erh tea.

Li *et al.* (2019) studied the volatile compounds of Sichuan Pingwu fuzhuan brick tea and other three fuzhuan teas from Guizhou, Hunan, and Shaanxi province by HP-SPME/GC-MS. The main volatile components belonged to aldehydes, ketones, and alcohols in decreasing order of volatile contents (%). The aldehydes were generated by oxidation and degradation of fatty acids and decarboxylation and oxidative deamination of amino acids (Ho *et al.* 2015). After postfermentation, the levels of hexanal, (E,Z)/(E,E)-2,4-heptadienal, (E)-2-pentenal were highly increased (Xu *et al.* 2007).

In some published articles in Chinese, the volatile composition of different dark teas was also compared. Zheng *et al.* (2018) described the aroma quality by descriptive words, for example, "stale" and "prolonged stale" aroma for qingzhuan tea of Hubei and ripened pu-erh tea. The volatiles with the highest content in qingzhuan tea are hexanal, nonanal, and β-ionone; linalool, methylbenzene, and 1,3-methoxy benzene are the main volatiles for fuzhuan tea, while linalool, α-cedarol, and limonene are the main compounds of liubao tea. Gu *et al.* (2009) also compared the volatiles of raw and ripened pu-erh tea by SDE, steam distillation–liquid/liquid extraction (SD-LLE), and Soxhlet extraction. The SDE showed the best extraction effects for volatiles, including benzyl alcohol, linalool, geraniol, and 1,2,3-trimethoxylbenzene. In postfermented pu-erh tea (ripened), the contents of 1,2,3-trimethoxylbenzene and 1,2,4-trimethoxylbenzene were about five- to tenfold those of raw pu-erh tea, but on the contrary, the contents of benzyl alcohol, phenethyl alcohol, and geraniol were highly decreased after postfermentation. Furthermore, E-nose can be also used in discriminating the pu-erh tea with different storage years (Yuan *et al.* 2019).

o obtain the best extraction and analytical results for methoxyphenolic compounds of pu-erh tea, Du *et al.* (2013) developed a method based on headspace solid-phase microextraction coupled with gas chromatography–mass spectrometry. CAR-PDMS fiber (carboxen/polydimethylsiloxane, 75 μm) had the best extraction efficiency for volatile methoxyphenolic compounds, and high temperature provided a better extraction. In the analysis of pu-erh tea samples, 1,2,3-trimethoxybenzene, 1,2,4-trimethoxybenzene, and 1,2-dimethoxy benzene were the major components, similar to many published articles. Although raw (crude, green) pu-erh tea was highly different from ripened pu-erh, the volatile characteristic compositions of green pu-erh tea were also studied. The total content of alcohols of green pu-erh tea is about 51%, which is about twofold that of other normal green tea, like longjing or biluochun green teas. Correspondingly, the ketones' level is much lower than in other green teas. Linalool, linalool oxide II/I, α-terpineol, and cedrol are the predominant alcohols in green pu-erh tea (Lv *et al.* 2015).

Ye *et al.* (2016) also compared the volatile constituents of ripened and crude pu-erh tea by E-nose and ultrasound-assisted extraction–dispersive liquid/liquid microextraction–gas chromatography–mass spectrometry. They also pointed that methoxyphenolic compounds accounted for 38.85% of total volatiles, compared with 10.11% in raw pu-erh tea. Linalool is the abundant compound in raw pu-erh tea, comprising about 24.5% of all volatiles, but highly decreased to 3.36% in ripened pu-erh tea. Different from other dark teas or other results on pu-erh tea, in this study, β-cyclocitral and safranal were the main aldehydes in raw pu-erh tea, and (E,E)2,4-heptadienal and β-cyclocitral were dominant in ripened pu-erh tea. According to other reports, floral and woody scent volatiles, including β-ionone, geranyl acetone, and α-ionone, had the highest levels in raw and ripened pu-erh tea. Similar research was recently published by Pang *et al.* (2019). The combination of gas chromatography olfactometry, aroma extract dilution analysis, odor activity values

(OAVs), and multivariate analysis revealed that the 19 odorants (OAV > 1) in ripened pu-erh tea are structurally and organoleptically similar. For example, 1,2,3-trimethoxybenzene, α-ionone, 1,2,3,4-tetramethoxybenzene, 1,2,4-trimethoxy benzene, and 1,2,3-trimethoxy-5-methylbenzene were the top five odorants in the order of OAVs, and four of those were methoxyphenolic compounds. However, in the raw pu-erh tea, α-ionone, linalool, 1,2,3-trimethoxybenzene, 2-methoxyphenol, and β-ionone were the top five odorants, in which α-ionone, linalool, and β-ionone have floral and woody scents. Therefore, the radar descriptive results showed that raw pu-erh tea had vector value on floral, refreshing, and sweet perceptions.

Xu *et al.* (2016) did a comprehensive comparison between raw and ripened pu-erh tea regarding aroma-active compounds with different scents. In total, 39 volatiles were identified as aroma-active compounds, from which most of them belonged to methoxyphenolic acids. The Osme value percentage of individual aroma-active compounds indicated that ripened pu-erh tea had a much higher relative aroma intensity percentage in stale/musty, woody scents, while raw pu-erh tea had a higher percentage in floral, fruity scents. Almost all studies indicated that the content of linalool decreased sharply after post-fermentation, which promoted the oxidation of linalool into linalool oxide I, II, III, IV. Some rose-like aroma compounds, including geraniol, nerolidol, and nerol, were detected in ripened pu-erh tea, but maybe their aroma was covered or interfered with by the stale/woody scent produced by methoxyphenolic compounds.

Except for the discriminative study for raw and ripened pu-erh tea, the brick tea could also be divided into two types. One is dark brick tea without Jin Hua (*Eurotium cristatum*); the other is fuzhuan brick tea with *Eurotium cristatum*, which is the predominate microorganism in fuzhuan teas (Huang *et al.* 2010). Nie *et al.* (2019) compared the volatile compounds between these two types of brick tea. In both teas, β-ionone had the largest OAV, but *Eurotium cristatum* increased the hydrolysis of β-primeverosides so that it produced more epoxydihydrolinalool I, epoxydihydrolinalool II, and methyl salicylate, while without *Eurotium cristatum*, the benzyl alcohol glycosides were converted into benzaldehyde and benzylacetate with high content in dark brick tea.

In addition to these reports, some research articles in Chinese also studied the aroma of various dark teas. HS-SPME is widely used in many studies because of its high efficiency in analysis. For example, Li *et al.* (2020) conducted a tracking analysis on a processed sample during manufacturing. During the *fahua*, which is the growing of *Eurotium cristatum*, and the biotransformation stage of tea's secondary metabolites, it is suggested that volatile compositions' varieties were increased and that the original volatiles, including longifolene, (+)-limonene, 3-carene, and other alkenes were gradually decreased from 0 to

22 days after postfermentation. Meanwhile, the alcohols, including linalool and its oxides, were increased significantly. Therefore, the aroma of fungal flower begins to appear after the growing of *Eurotium cristatum* but mainly depended on the production of alcohols like linalool and its oxides.

Zhao *et al.* (2017) compared the aroma components and quality of fuzhuan tea made with raw dark teas from different regions. The fuzhuan tea produced in Hunan, Shanxi, and Zhejiang provinces all presented the main volatiles as alcohols, aldehydes, and ketones. Compared with the origin production place (Hunan) of fuzhuan tea, the fuzhuan tea from Zhejiang showed a grassy scent, while the Shanxi sample exhibited a roasting scent.

Shen *et al.* (2017) also focused on the aroma analysis of Hunan fuzhuan tea and reported that alcohols, acids, and ketones accounted for 23%, 18%, and 11% of total volatiles, respectively. In this study, volatiles were extracted by SDE, so that many low-boiling-point aldehydes maybe not be well collected after long-term brewing. On the other hand, palmitic acid was preserved for its properties and detected as the highest content at 17.29% average content.

Aroma quality is also varied with the storage years, like red wine and raw pu-erh tea. Li *et al.* (2016) made a descriptive quality analysis for different fuzhuan teas from four-year to 54-year storage times. As storage was prolonged, the floral scent gradually disappeared, but the woody and Chinese medicinal scent appeared. Although there has been no volatiles analysis for the Chinese medicinal scent, it may be related to borneol or its derivatives. With the storage time, the relative content of acids was correspondingly increased, but the aldehydes and ketones were gradually decreased. It was interesting that the typical floral compound linalool was not detected by SDE in the essential oil of fuzhuan tea.

Qingzhuan tea is a kind of dark tea mainly produced in Chibi, Hubei province of China. It was reported that the volatile compounds of qingzhuan tea were 2,6-di-butyl-p-cresol (16.26%) and cypressneol (14.82%) as relative content (Yuan *et al.* 2014).

Because the rapid growing of microorganisms during pile-fermentation of postfermentation, it was suggested that the aroma formation is tightly related to the predominant fungus involved. We previously did a tracking analysis on fermented tea samples and found that the aroma compounds could be enhanced at the beginning stage. However, after long-term fermentation, almost all odor-active compounds disappeared (Cao *et al.* 2018). The formation of the aroma quality of dark tea during pile-fermentation based on multi-omics was also studied, which demonstrated a tight correlation between microbial communities and volatile compounds ($|r| > 0.7$, $p < 0.05$). *Candida*, *Debaryomyces*, *Cyberlindnera*, and *Penicillium* had high positive correlations with alcohols and ketones. *Byssochlamys* had a high positive correlation with ketones, alkenes, and esters, and *Rasamsonia* had negative correlation with alcohols, ketones, and aldehydes. *Thermoascus* and *Thermomyces* had a positive correlation with aldehydes (Hu *et al.* 2021).

About the formation mechanism of aroma compounds of dark tea, most were similar to those of green tea and black tea (Ho, Zheng, and Li 2015), but the microbial fermentation may endow some specific formation pathway, such as the degradation and oxidation of catechins to produce methoxyphenolic acids. The decrease of galloylated catechin produce high content of gallic acid at the beginning stage of postfermentation, but after long-term fermentation, gallic acid's level decreased (Qin *et al.* 2012). Some nonvolatile compounds may be formed by microorganisms like *Aspergillus niger*, *Aspergillus fumigatus*, *Aspergillus flavus*, etc., but the volatiles' formation by microorganisms is still not clear (Wang *et al.* 2014). As the main floral scent volatiles, geraniol, linalool, and their oxides, were mainly biosynthesized in the tea plant by geraniol synthase and linalool synthase from the geranyl-pyrophosphate (geranyl-PP) and were stored as glycosidic precursors that could be further hydrolyzed by glucosidase and primeverosidase (Nishikitani *et al.* 1999; Wang *et al.* 2000). The ketones are also main aroma contributors for many dark teas. They include β-ionone, α-ionone, β-damascone, geranylacetone, and theaspirone, which have been described as flowery, woody, violet, and floral scented. They are derived from the oxidation of β-carotene and phytofluene (Schuh and Schieberle 2006). Lipids, especially long-chain unsaturated fatty acids, are also important precursors for many aldehydes including hexanal, pentanal, (Z)-3-hexenal, (Z)-4-heptanal, and (E,Z)-2,6-nonadienal. These compounds usually show grassy, green, cucumber-like, and green scents. As the typical stale-like compounds, methoxyphenolic compounds were supposed to be produced by the degradation of tea polyphenols. It was interesting that some esters were detected as main volatiles in some dark teas. So far, esters, especially low-molecular-weight esters, are critical aroma compounds in Chinese *baijiu* (wine), but in dark tea, the production of esters might have a mechanism similar to that observed in wine fermentation, in which the *Saccharomyces cerevisiae* converts carbohydrates into ethyl acetate (Fan *et al.* 2019). In pu-erh tea, *Pichia farinose* and *Arxula adeninivorans* have been identified (Zhang *et al.* 2013).

Different from green tea or black tea aroma, dark tea's aroma is not well accepted by many traditional consumers, perhaps because of the stale and woody aroma. Therefore, some dark tea fermented by a single fungus has been developed. Furthermore, the aroma extraction methods also have different effects on volatile enrichment, which is also like the brewing of many dark teas. For green tea, the brewed water is about 90°C, but dark tea brewing may need boiling water, even continuously heated by boiling in teaware. The low-boiling off-odor compounds could be evaporated, allowing the esters or other floral compounds to make a better flavor for brewed dark tea infusion.

9.4 PERSPECTIVES

To reveal the crucial aroma compounds responsible for the typical flavor of various dark teas, the techniques and methods of molecular sensory science should be applied. Although methoxyphenolic compounds were widely recognized as the typical aroma compounds of dark teas, most of their presence could be observed in dark teas only with a high degree of postfermentation processing, as the degradation products of its precursor, galloylated catechins. However, many dark teas still have a high content of polyphenols; therefore the aroma compounds may be highly different from those of ripened pu-erh tea, a typical type of deep postfermentation. The microbial transformation of aroma formation also needs further exploration.

REFERENCES

Cao, L., X. Guo, G. Liu *et al.* 2018. A comparative analysis for the volatile compounds of various Chinese dark teas using combinatory metabolomics and fungal solid-state fermentation. *Journal of Food and Drug Analysis* 26:112–123.

Du, L., J. Li, W. Li, Y. Li, T. Li and D. Xiao. 2014. Characterization of volatile compounds of pu-erh tea using solid-phase microextraction and simultaneous distillation—extraction coupled with gas chromatography—mass spectrometry. *Food Research International* 57:61–70.

Du, L., C. Wang, J. Li, D. Xiao, C. Li and Y. Xu. 2013. Optimization of headspace solid-phase microextraction coupled with gas chromatography-mass spectrometry for detecting methoxyphenolic compounds in Pu-erh tea. *Journal of Agricultural and Food Chemistry* 61:561–568.

Du, L., C. Wang, C. Zhang, L. Ma, Y. Xu and D. Xiao. 2019. Characterization of the volatile and sensory profile of instant Pu-erh tea using GC × GC-TOFMS and descriptive sensory analysis. *Microchemical Journal* 146:986–996.

Fan, G., C. Teng, D. Xu *et al.* 2019. Enhanced production of ethyl acetate using co-culture of *Wickerhamomyces anomalus* and *Saccharomyces cerevisiae*. *Journal of Bioscience and Bioengineering* 128:564–570.

Gu, X., Z. Zhang, X. Wan, J. Ning, C. Yao and W. Shao. 2009. Simultaneous distillation extraction of some volatile flavor components from pu-erh tea samples—comparison with steam distillation-liquid/liquid extraction and soxhlet extraction. *International Journal of Analytical Chemistry:* 276713.

Ho, C. T., X. Zheng, and S. Li. 2015. Tea aroma formation. *Food Science and Human Wellness* 4:9–27.

Hu, S., C. He, Y. Li *et al.* 2021. The formation of aroma quality of dark tea during pile-fermentation based on multi-omics. *LWT* 47:111491.

Huang, Y. H., Chen, J. H., Zhou, Y., and Chen, X. Y. 2010. Differences in sensory quality and chemical composition of Fuzhuan Tea of different storage ages. *Food Science* 31:228–232.

Li, J., Y. Xu, M. Chen, G. Deng, K. Wu, and L. Jiang. 2020. Analysis of changes in volatile components during processing of handmade Fuzhuan tea. *Food Science* 41:144–154 (in Chinese with English abstract).

Li, M., Y. Xiao, K. Zhong, J. Bai, and H. Gao. 2019. Characteristics and chemical compositions of Pingwu Fuzhuan brick-tea, a distinctive post-fermentation tea in Sichuan province of China. *International Journal of Food Properties* 22:878–889.

Li, S., Y. Shen, D. Fu, Z. Liu and J. Huang. (2016). Organoleptic quality analysis of Fuzhuan brick teas in different storage years. *Tea Science* 5:500–504.

Li, Y., Y. Huang, X. Liu *et al*. 2018. Comparative analysis of volatile components in Fu brick tea by simultaneous distillation and headspace solid-phase micro extraction. *Science and Technology of Food Industry* 39:246–252.

Lv, S., Y. Wu, Y. Song *et al*. 2015. Multivariate analysis based on GC-MS fingerprint and volatile composition for the quality evaluation of Pu-erh green tea. *Food Analytical Methods* 8:321–333.

Lv, S., Y. Wu, J. Zhou *et al*. 2014. The study of fingerprint characteristics of Dayi Pu-erh tea using a fully automatic HS-SPME/GC—MS and combined chemometrics method. *PLoS One* 9:e116428.

Nie, C., X. Zhong, L. He *et al*. 2019. Comparison of different aroma-active compounds of Sichuan Dark brick tea (*Camellia sinensis*) and Sichuan Fuzhuan brick tea using gas chromatography—mass spectrometry (GC—MS) and aroma descriptive profile tests. *European Food Research and Technology* 245:1963–1979.

Nishikitani, M., D. Wang, K. Kubota, A. Kobayashi and F. Sugawara. 1999. (Z)-3-hexenyl and trans-linalool 3,7-oxide beta-primeverosides isolated as aroma precursors from leaves of a green tea cultivar. *Bioscience Biotechnology and Biochemistry* 63:1631–1633.

Pang, X., W. Yu, C. Cao *et al*. 2019. Comparison of potent odorants in raw and ripened Pu-erh tea infusions based on odor activity value calculation and multivariate analysis: Understanding the role of pile fermentation. *Journal of Agricultural and Food Chemistry* 67:13139–13149.

Qin, J. H., N. Li, P. F. Tu, Z. Z. Ma, and L. Zhang. 2012. Change in tea polyphenol and purine alkaloid composition during solid-state fungal fermentation of postfermented tea. *Journal of Agricultural and Food Chemistry* 60:1213–1217.

Schuh, C., and P. Schieberle. 2006. Characterization of the key aroma compounds in the beverage prepared from darjeeling black tea: Quantitative differences between tea leaves and infusion. *Journal of Agricultural and Food Chemistry* 54:916–924.

Shen, C., Y. Deng, Y. Zhou *et al*. 2017. Research of quality features and aroma components in Hunan Fu brick tea. *Journal of Tea Science* 37:38–48.

Shi, J., Y. Zhu, Y. Zhang, Z. Lin, and H. P. Lv. 2019. Volatile composition of Fu-brick tea and Pu-erh tea analyzed by comprehensive two-dimensional gas chromatography-time-of-flight mass spectrometry. *LWT* 103:27–33.

Wang, D., T. Yoshimura, K. Kubota, and A. Kobayashi. 2000. Analysis of glycosidically bound aroma precursors in tea leaves. 1. Qualitative and quantitative analyses of glycosides with aglycons as aroma compounds. *Journal of Agricultural and Food Chemistry* 48:5411–5418.

Wang, W., L. Zhang, S. Wang *et al*. 2014.8-C N-ethyl-2-pyrrolidinone substituted flavan-3-ols as the marker compounds of Chinese dark teas formed in the post-fermentation process provide significant antioxidative activity. *Food Chemistry* 152:539–545.

Xu, X., H. Mo, M. Yan, and Y. Zhu. 2007. Analysis of characteristic aroma of fungal fermented Fuzhuan brick-tea by gas chromatography/mass spectrophotometry. *Journal of the Science of Food and Agriculture* 87:1502–1504.

Xu, Y. Q., C. Wang, C. W. Li *et al.* 2016. Characterization of aroma-active compounds of Pu-erh Tea by headspace solid-phase microextraction (HS-SPME) and simultaneous distillation-extraction (SDE) coupled with GC-olfactometry and GC-MS. *Food Analytical Methods* 9:1188–1198.

Ye, J., W. Wang, C. Ho, J. Li, X. Guo, M. Zhao *et al.* 2016. Differentiation of two types of pu-erh teas by using an electronic nose and ultrasound-assisted extraction-dispersive liquid—liquid microextraction-gas chromatography-mass spectrometry. *Analytical Methods* 8:593–604.

Yuan, H., X. Chen, Y. Shao, Y. Cheng, and Z. Wu. 2019. Quality evaluation of green and dark tea grade using electronic nose and multivariate statistical analysis. *Journal of Food Science* 84:3411–3417.

Yuan, S., Z. Bai, Y. Huang, X. Lai, C. Wu, and W. Zhao. 2014. Analysis of aroma components in three kinds of dark tea. *Food Science* 35:252–256.

Zhang, L., Z. Zhang, Y. Zhou, T. Ling, and X. Wan. 2013. Chinese dark teas: Postfermentation, chemistry and biological activities. *Food Research International* 53:600–607.

Zhao, R., D. Wu, Y. Jiang, and Q. Zhu. 2017. Analysis on the aroma components and quality of Fu brick tea made with raw dark teas from different regions. *Journal of Human Agricultural University (Natural Sciences)* 43:551–555.

Zheng, P., P. Liu, S. Wang *et al.* 2018. Comparative analysis of the aroma components in five kinds of dark tea. *Science and Technology of Food Industry* 22:82–86.

Index

Note: Page numbers in *italics* indicate a figure and page numbers in **bold** indicate a table on the corresponding page.

For Product Safety Concerns and Information please contact our EU
representative GPSR@taylorandfrancis.com
Taylor & Francis Verlag GmbH, Kaufingerstraße 24, 80331 München, Germany

www.ingramcontent.com/pod-product-compliance
Lightning Source LLC
Chambersburg PA
CBHW060400220326
41598CB00023B/2977

9 7 8 1 0 3 2 1 6 1 9 3 8